TOPICS IN ALGEBRAIC AND ANALYTIC GEOMETRY

Notes from a course of

PHILLIP GRIFFITHS

Written and revised by

JOHN ADAMS

PRINCETON UNIVERSITY PRESS

AND

UNIVERSITY OF TOKYO PRESS

PRINCETON, NEW JERSEY

1974

Published by Princeton University Press, Princeton and London

L.C. Card: 74-2968

I.S.B.N.: 0-691-08151-4

Library of Congress Cataloging in Publication Data
will be found on the last printed page of this book

Published in Japan exclusively
by University of Tokyo Press
in other parts of the world by
Princeton University Press

Printed in the United States of America

Introduction

This is a revised version of the notes taken from a class taught at Princeton University in 1971-1972. The table of contents gives a good description of the material covered.

The notes focus on comparison theorems between the algebraic, analytic, and continuous categories.

CONTENTS

Chapter One

§I Sheaf theory, ringed spaces

On $\mathbb{C}^n$, the space of n complex variables, there is a sequence of sheaves defined in the usual topology:

$$O_{cont} \supset O_{diff} \supset O_{hol} \supset O_{alg}$$

where for an open set $U \subset \mathbb{C}^n$:

$\Gamma(U, O_{cont})$ = continuous complex-valued functions defined on U.

$\Gamma(U, O_{diff})$ = C^∞ complex-valued functions defined on U.

$\Gamma(U, O_{hol})$ = holomorphic functions defined on U.

$\Gamma(U, O_{alg})$ = rational holomorphic functions defined on U.

(A holomorphic function φ on U is said to be rational just in case each a in U has a neighborhood W in U such that there are two polynomials p, q with q nowhere zero in W, with $\varphi = p/q$ in W. In fact, because the polynomial ring $\mathbb{C}[z_1, \ldots, z_n]$ is a unique factorization domain, one can always take $W = U$.)

If W is an open set properly contained in the open set U then there is always an element of $\Gamma(W, O_{cont})$ which does not extend to U, and similarly for O_{diff}. But O_{hol} and O_{alg} behave differently: We will prove a little theorem about this.

I.A THEOREM (Hartog's removable singularities theorem) In case $n \geq 2$ and $W = U$ less a point, every holomorphic function on W extends to U.

We may suppose that $W = U - \{(0, \ldots, 0)\}$. If δ is a small positive number such that $\Delta^n(\delta) = \{z : \sup_{i=1,\ldots,n} |z_i| \leq \delta\}$ is contained in U we define for each f holomorphic in W

$$f_1(z_1, \ldots, z_n) = \frac{1}{2\pi\sqrt{-1}} \int_{|z_1|=\delta} \frac{f(\xi, z_2, \ldots, z_n)}{\xi - z_1} \, d\xi$$

for z in the interior of the polydisk. It will suffice to show that $f = f_1$ in the interior of the polydisk less its center. But for a point inside the polydisk with $z_2 \neq 0$ the formula

$$f(z_1, \ldots, z_n) = \frac{1}{2\pi\sqrt{-1}} \int_{|z_1|=\delta} \frac{f(\xi, z_2, \ldots, z_n)}{\xi - z_1} \, d\xi$$

is valid, so $f = f_1$ where both are defined.

This type of behavior is more pronounced in the case of O_{alg}. Every holomorphic rational function on W will extend to U unless there is a polynomial with zeroes in U but no zeroes in W.

The sheaf O_{alg} may naturally be restricted to a coarser topology on $\mathbb{C}^n$, the Zariski topology. A set in $\mathbb{C}^n$ is a Zariski closed set in case it is the locus of zeroes of a set of polynomials— one can always take the set of polynomials to be an ideal in the polynomial ring. The Zariski closed set associated to an ideal I will be denoted $V(I)$ ("variety of I"). The complement of a Zariski closed set is a

Zariski open set. This defines a topology on $\mathbb{C}^n$, coarser than the usual topology. Two algebraic facts about this topology are

(1) $V(I) = \phi$ just in case $I = \mathbb{C}[z_1, \ldots, z_n]$

(2) $V(I) = V(J)$ just in case $\mathrm{rad}(I) = \mathrm{rad}(J)$.

The first of these facts is called the Hilbert Nullstellensatz. For a proof of this, and the deduction of (2) from (1), see Safarevic [35], or Lefschetz [20].

Henceforth when we consider the sheaf O_{alg} on $\mathbb{C}^n$ it will usually be with respect to the Zariski topology, sometimes denoted $\mathbb{C}^n_{Zar}$.

We have met with four examples $(\mathbb{C}^n, O_{cont})$, $(\mathbb{C}^n, O_{diff})$, $(\mathbb{C}^n, O_{hol})$, $(\mathbb{C}^n_{Zar}, O_{alg})$ of a topological space with a sheaf of rings. For any topological space X there is another example, $(X, O_{\mathbb{C}})$, where $O_{\mathbb{C}}$ is the sheaf of complex-valued (not necessarily continuous) functions.

A topological space with a sheaf of rings on it, (X, O), such that O is a subsheaf of the sheaf $O_{\mathbb{C}}$, is called a <u>ringed space</u>.

A <u>morphism of ringed spaces</u> (X, O_X) and (Y, O_Y) is a continuous map $\varphi : X \longrightarrow Y$ such that, for any open set $U \subset Y$, and $f \in \Gamma(U, O_Y)$, $f \circ \varphi \in \Gamma(\varphi^{-1}(U), O_X)$. An isomorphism of ringed spaces (X, O_X) and (Y, O_Y) is a homeomorphism of $\varphi : X \longrightarrow Y$ such that $\Gamma(\varphi^{-1}(U), O_X) = \Gamma(U, O_Y)$ for all open U in Y.

---

* A ringed space is sometimes more generally defined as simply a topological space with a sheaf of rings. Then the definition of morphism is more complicated.

If U is an open subset on the ringed space (X, O_X), then $O_X|_U$ makes U into a ringed space so that the natural inclusion $U \longrightarrow X$ induces a morphism of ringed spaces.

The terminology of ringed spaces provides a convenient, general way to speak of the class of distinguished functions singled out on a topological space by some special structure on that space. For example, we could define a continuous or differentiable manifold to be a Hausdorff ringed space (X, O) such that each point has an open neighborhood which is isomorphic, with its induced ringed space structure, to either $(\mathbb{R}^n, O_{cont})$ or $\mathbb{R}^n, O_{diff})$. Then a morphism between the ringed spaces associated to two differentiable manifolds, $(X, O_{diff}) \xrightarrow{\varphi} (Y, O_{diff})$, is the same thing as a differentiable map in the usual sense.

We define a <u>complex manifold</u> as a Hausdorff ringed space, (X, O), such that each point has a neighborhood which is isomorphic to (U, O_{hol}), with U an open subset in $\mathbb{C}^n$. A morphism of ringed spaces between two complex manifolds is called a holomorphic mapping.

The stalk of an arbitrary point a in a ringed space (X, O) is denoted O_a. In the examples of $(\mathbb{C}^n, O_{cont})$, $(\mathbb{C}^n, O_{diff})$, $(\mathbb{C}^n, O_{hol})$, $(\mathbb{C}^n_{Zar}, O_{alg})$ the stalk of a point is always a local ring. For example, the stalk of the origin in $(\mathbb{C}^n, O_{hol})$ is the ring $\mathbb{C}<z_1, \ldots, z_n>$ of convergent power series in n variables; the stalk of the origin in $(\mathbb{C}^n_{Zar}, O_{alg})$ is the localization of the polynomial ring $\mathbb{C}[z_1, \ldots, z_n]$ at the ideal $(z_1, \ldots, z_n)$. We shall be dealing with such <u>local ringed spaces</u> almost exclusively. A ringed space is local just in case it has the properties

(1) A distinguished function which is non-zero at some point is also non-zero in some neighborhood of the point.

(2) The reciprocal of a non-zero distinguished function defined in a neighborhood of a point is also distinguished over a perhaps smaller neighborhood.

If $(X, \mathcal{O})$ is a local ringed space and $f \in \Gamma(X, \mathcal{O})$ then $\{x \in X : f(x) = 0\}$ is closed in X.

The chief motivation for introducing the terminology of ringed spaces is in the study of analytic and algebraic sets. An <u>algebraic set</u> in $\mathbb{C}^n$ is just a Zariski closed set. An <u>analytic set</u> in $\mathbb{C}^n$ is characterized by being locally the locus of zeroes of some finite number of holomorphic functions - that is, there is a covering $\{U_i\}$ of $\mathbb{C}^n$ such that the intersection of the set with U_i is $\{x \in U_i : f_{ij}(x) = 0, j = 1, \ldots, N\}$ for some $f_{ij} \in \Gamma(U_i\ \mathcal{O}_{hol})$. If X is an algebraic (resp. analytic) set in $\mathbb{C}^n$, one can form a sheaf of ideals on the ringed space $(\mathbb{C}^n_{Zar}, \mathcal{O}_{alg})$ (resp. $(\mathbb{C}^n, \mathcal{O}_{hol})$) by $I_X(U) = \{f \in \Gamma(U, \mathcal{O}) : f = 0 \text{ on } X \cap U\}$. The quotient sheaf $\mathcal{O}/I_X$ will be a sheaf of rings supported on the closed subspace X. It will be a subsheaf of the sheaf of complex functions, so this makes X into a ringed space. Thus any algebraic (resp. analytic) subset of $\mathbb{C}^n$ has a natural ringed space structure - in fact a local ringed space structure, since it arises by taking a quotient from a sheaf with local rings for stalks.

The same procedure can be followed to define analytic subsets of open sets in $\mathbb{C}^n$, and to show how any such analytic set has a natural ringed space structure. We could also do this in the algebraic case, but in this case we would get nothing new, as we shall see later.

An algebraic variety is by definition a ringed space which is locally isomorphic to the ringed space defined by an algebraic subset of $\mathbb{C}^n$. An analytic space, or analytic variety, is a ringed space which is locally isomorphic to the ringed space defined by an analytic subset of an open set in $\mathbb{C}^n$. A morphism of two algebraic varieties (or holomorphic rational mapping) is a ringed space morphism; similarly, for the morphisms of analytic spaces, which are called holomorphic mappings.

When working with analytic spaces it is usually helpful to require them to be Hausdorff spaces if possible. There is an analogous condition which we can put on algebraic varieties. First note that, given two algebraic varieties (or analytic spaces) X and Y, one can form their product as an algebraic variety (or analytic space), $X \times Y$: If X and Y were algebraic subsets of $\mathbb{C}^n$ this construction would be clear, and the general case reduces to this and patching. In the analytic case one reduces to X, Y analytic subsets of open sets in $\mathbb{C}^n$. For a proof in the algebraic case, see Serre [27]. For an analytic space X, the condition of Hausdorffness is equivalent to the diagonal map $X \xrightarrow{\Delta} X \times X$ being closed. If Y is an algebraic variety one says that Y is separated just in case the diagonal map $X \xrightarrow{\Delta} Y \times Y$ is closed (in the Zariski topology on $Y \times Y$, which is not the product topology).

Algebraic subvarieties of algebraic varieties, and analytic subsets of analytic spaces are defined by analogy with $\mathbb{C}^n$, and given ringed space structure. Any algebraic subvariety of a separated algebraic variety is itself separated, just as any subspace of a Hausdorff space is Hausdorff. $\mathbb{C}^n_{Zar}$ is separated. From now on we will assume that all our analytic spaces are Haudorff and all our algebraic varieties are separated.

We will consider some examples of analytic and algebraic subsets of $\mathbb{C}^2$. A polynomial in two variables $p(z_1, z_2)$ defines an algebraic subset of $\mathbb{C}^2$, $V(p(z_1, z_2))$

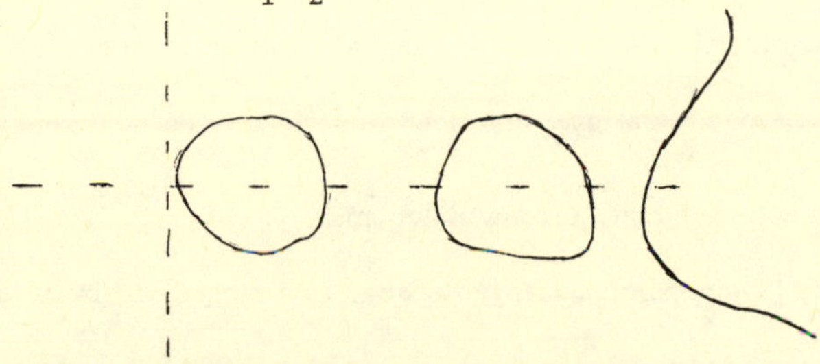

$V(z_2^2 - (z_2)(z_2 - 1)(z_2 - 2)(z_2 - 3)(z_2 - 4))$

A pair of polynomials in two variables, $f(z_1, z_2)$ and $g(z_1, z_2)$ will, providing that neither is constant and they have no common factor, define a finite set of points. Any algebraic subset of $\mathbb{C}^2$ is either like the first example or the second, or a union of both. See Safarevic [35] or Lefschetz [20].

Let X be an algebraic subset of $\mathbb{C}^2$. We propose to describe the sheaf of ideals I_X on $(\mathbb{C}^2_{Zar}, O_{alg})$. For any $f \in \mathbb{C}[z_1, z_2]$ we denote the Zariski open set $\mathbb{C}^2 - V(f)$ by $D(f)$. Then $\Gamma(D(f), O_{alg}) = \mathbb{C}[z_1, z_2]_f = \mathbb{C}[z_1, z_2, z_3]/(1 - z_3 f)$. Now an element g/f^n of $\Gamma(D(f))$ will vanish on X just in case g vanishes on X, and this implies that $I_X(D(f)) = I_X(\mathbb{C}^2)\Gamma(D(f))$. Since the sets $D(f)$ form a basis for the Zariski topology on $\mathbb{C}^2$, we find that the sheaf of ideals I_X is generated over O_{alg} by its global sections.

As this discussion suggests, the algebraic subsets of the Zariski open subsets of $\mathbb{C}^n$ are exactly the Zariski closed subsets, that is, the closed subsets inherited from the Zariski topology on $\mathbb{C}^n$. For if $\mathbb{C}^n - V(I) = U$ is a Zariski open then $\Gamma(U, O_{alg}) = \mathbb{C}[z_1, \ldots, z_n]$ localized by the multiplicative subset I. If we define an algebraic subset of U to be the locus of zeroes of an ideal in $\Gamma(U, O_{alg})$, then there will be an ideal $J \subset \mathbb{C}[z_1, \ldots, z_n]$ such that $I = J\Gamma(U, O_{alg})$, so $V(I) = V(J) \cap U$. This also shows that a subset of $\mathbb{C}^n$ which is locally given (in Zariski opens) as the zeroes of rational functions is a globally defined algebraic set.

Another point to notice is that, for any polynomial $f \in \mathbb{C}[z_1, \ldots, z_2]$, $D(f) = \mathbb{C}^n_{Zar} - V(I)$, is isomorphic, as ringed space, to a closed algebraic subset of $\mathbb{C}^{n+1}$: It will be isomorphic to $V(1 - z_{n+1} f)$. This shows that any open subset of an algebraic variety is an algebraic variety.

As an example of an analytic subset of $\mathbb{C}^2$, consider the set $X = V(z_2 - e^{z_1})$

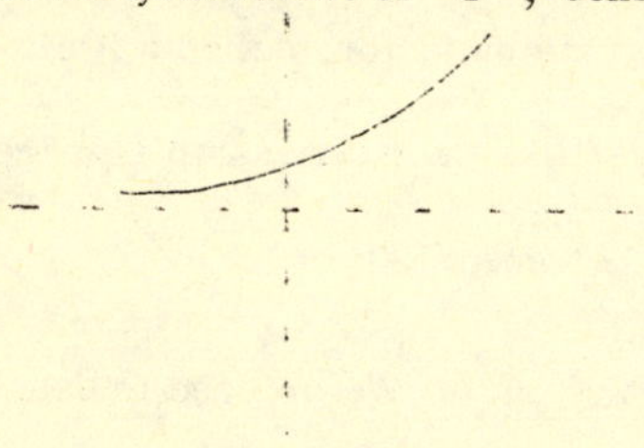

This set will not be algebraic, and as a ringed space will be isomorphic to $(\mathbb{C}, O_{hol})$.

Chapter One

§2 Local structure of analytic and algebraic sets

A discussion of the local properties of analytic subsets of open sets in $\mathbb{C}^n$ requires some preparation. We shall mostly limit our discussion to the case of analytic varieties defined by a single equation. What we need is the

I. B THEOREM (Weierstrass preparation and division theorems)

Let f be a holomorphic function defined in a neighborhood of the closed polydisk $\overline{\Delta^n(r)}$. Suppose that the function of one variable defined by $y(z_n) = f(0, \ldots, 0, z_n)$ in a neighborhood of $\overline{\Delta^1(r)}$ is not identically zero and has a zero of order p at 0. Suppose also that $f(z_1, \ldots, z_n)$ is never zero when $|z_n| = r$, $(z_1, \ldots, z_{n-1}) \in \Delta^{n-1}(r)$. Then

(1) There is a holomorphic function u which is a unit in $\Delta^n(r)$, such that both it and its inverse are bounded on the polydisk, and there are bounded holomorphic functions

$$a_0(z_1, \ldots, z_{n-1}), \ldots, a_{p-1}(z_1, \ldots, z_{n-1})$$

on $\Delta^{n-1}(r)$ such that $a_i(0, \ldots, 0) = 0$ and

$$f = u(z_n^p + a_{p-1}(z_1, \ldots, z_{n-1})z_n^{p-1} + \ldots + a_0(z_1, \ldots, z_{n-1}))$$

on the polydisk.

(2) Given φ holomorphic on $\Delta^n(r)$, there is a function $a(z_1, \ldots, z_n)$ holomorphic on $\Delta^n(r)$ and functions $b_0(z_1, \ldots, z_{n-1}), \ldots, b_{p-1}(z_1, \ldots, z_{n-1})$ holomorphic on $\Delta^{n-1}(r)$ such that

$$\varphi = af + z_n^{p-1} b_{p-1}(z_1, \ldots, z_{n-1}) + \ldots + b_0(z_1, \ldots, z_{n-1})$$

<u>on</u> $\Delta^n(r)$. <u>There is a constant</u> M , <u>independent of</u> φ, <u>such that</u>

$$\|a\|_{\Delta^n(r)},\ \|b_i\|_{\Delta^{n-1}(r)} \leq M\,\|\varphi\|_{\Delta^n(r)}$$

(1) As $(z_1,\ldots,z_{n-1})$ varies in $\Delta^{n-1}(r)$ we get a holomorphically varying family of holomorphic functions of z_n , each defined in a neighborhood of $\overline{\Delta^1(r)}$. We'll use the notation (z',z_n) for a point of $\Delta^n(r) = \Delta^{n-1}(r) \times \Delta(r)$. Now

$$b_0(z') = \frac{1}{2\pi\sqrt{-1}} \int_{|z'_n|=r} \frac{\partial f}{\partial z_n}(z',z_n)\,\frac{dz_n}{f(z',z_n)}$$

$$= \text{number of zeroes of } f(z',z_n) \text{ in } |z_n| \leq r$$

$$= p \quad \text{identically, by continuity} .$$

Denote by $t_1(z'),\ldots,t_p(z')$ the aforementioned zeroes. The symmetric function $\sigma_k(z') = (t_1(z'))^k + \ldots + (t_p(z'))^k$ admits the representation

$$\sigma_k(z') = \frac{1}{2\pi\sqrt{-1}} \int_{|z_n|=r} z_n^k\,\frac{\partial f}{\partial z_n}(z',z_n)\,\frac{dz_n}{f(z^1,z_n)}$$

and is thus a holomorphic function of z' . The rth elementary symmetric function of the $t_j(z)$

$$a_{p-r}(z') = \sum_{i_1<\ldots<i_p} t_{i_1}(z')\ldots t_{i_p}(z')$$

can be represented as a polynomial with rational coefficients in the $\sigma_k(z')$, hence is holomorphic in z'. Setting

$$\pi(z_1, \ldots, z_n) = z_n^p + a_{p-1}(z') z_n^{p-1} + \ldots + a_0(z')$$

it remains to prove that f/π is holomorphic on $\Delta^n(r)$. Note that f/π is defined and non-vanishing, because for each z' $f(z', z_n)$ and $\pi(z', z_n)$ are holomorphic functions of z_n with the same zeroes. Note also that we could have done all this on a slightly larger polydisk, so we can assume that f/π is defined on a neighborhood of $\overline{\Delta^n(r)}$. Then we have the representation in $\Delta^n(r)$

$$\frac{f}{\pi}(z', x) = \frac{1}{2\pi\sqrt{-1}} \int_{|z_n|=r} \frac{f(z', z_n)}{\pi(z', z_n)} \frac{dz_n}{z_n - x}$$

which shows that f/π is holomorphic.

Since everything could have been done on a slightly larger polycylinder, u and the a_i's must be bounded.

(2) First choose $0 < r' < r$ so that $\pi(z', z_n)$ has no zeroes in $\Delta^{n-1}(r) \times \Delta(r)$ with $|z_n| \geq r'$. Given φ, set

$$a(z', x) = \frac{1}{2\pi\sqrt{-1}} \int_{|z_n|=r'} \frac{\varphi(z', z_n)}{\pi(z', z_n)} \frac{1}{z_n - x} dz_n$$

which will be holomorphic in $\Delta^{n-1}(r) \times \Delta(r')$ and will not change if r' increases.

$$\varphi(z',x) - a(z',x)\pi(z',x) = \frac{1}{2\pi\sqrt{-1}} \int_{|z_n|=r'} \varphi(z',z_n)\left\{1-\frac{\pi(z',x)}{\pi(z',z_n)}\right\}\frac{dz_n}{z_n-x}$$

$$= \frac{1}{2\pi\sqrt{-1}} \int_{|z_n|=r'} \frac{\varphi(z',z_n)}{\pi(z',z_n)}\left\{z_n^p - x^p + a_{p-1}(z')(z_n^{p-1} - x^{p-1}) + \dots + a_1(z')(z_n - x)\right\}\frac{dz_n}{z_n-x}$$

$$= \frac{1}{2\pi\sqrt{-1}} \int_{|z_n|=r'} \frac{\varphi(z',z_n)}{\pi(z',z_n)}\left\{c_{p-1}(z',z_n)x^{p-1} + \dots + c_0(z',z_n)\right\} dz_n$$

where $c_i(z',z_n)$ is a polynomial in z_n, of degree less than p, with coefficients linear combinations of the a_i. Let

$$b_i(z') = \frac{1}{2\pi\sqrt{-1}} \int_{|z_n|=r'} \frac{\varphi(z',z_n)}{\pi(z',z_n)}\left\{c_i(z',z_n)\right\} dz_n \quad .$$

Since $\pi(z',z_n)$ is bounded away from 0 on $|z_n| = r'$, and since the a_i are bounded, there is M such that

$$\|b_i\|_{\Delta^{n-1}(r)} \leq M \|\varphi\|_{\Delta^n(r)}$$

with M independent of φ. On $|z_n| = r'$

$$|a(z',z_n)| = \frac{|\varphi - (b_{p-1}z_n^{p-1} + \dots + b_0)|}{|\pi(z',z_n)|}$$

so for some M' independent of φ,

$$|a(z',z_n)| \leq \frac{M' \|\varphi\|_{\Delta^n(r)}}{\inf_{|z_n|\geq r'} |\pi(z',z_n)|}$$

Letting r' go to r and using the maximum principle, we get

$$\|a\|_{\Delta^n(r)} \leq M'' \|\varphi\|_{\Delta^n(r)}$$

for some M'' independent of φ.

A polynomial $z_n^p + a_{p-1}(z_1, \ldots, z_{n-1})z_n^{p-1} + \ldots + a_0(z_1, \ldots, z_{n-1})$ with $a_i(0, \ldots, 0) = 0$ is called a distinguished pseudo-polynomial.

Note that, given a holomorphic function defined in a neighborhood of the origin, there is a linear change of coordinates such that the hypotheses of the theorem are satisfied on small enough polydisks.

The Weierstrass preparation and division theorems suffice to establish some basic facts about the local rings of holomorphic functions of points in $\mathbb{C}^n$.

I.C PROPOSITION The local ring O_0 of the origin in $\mathbb{C}^n$ is noetherian.

The proof is by induction on n, the $n=0$ being obvious. Given an ideal I of O_0, pick $f \in I$, $f \neq 0$. After a possible change of coordinates we may assume that f is a distinguished pseudo-polynomial in z_n over the local ring in $n-1$ variables, O'_0. Let $I' = I \cap O'_0[z_n]$. $O'_0[z_n]$ is noetherian by induction and the Hilbert basis theorem. Suppose then that $f_1, \ldots, f_n$ is a basis for I'. Given $\varphi \in I$, use the division theorem to write $\varphi = af + b$, b an element of $O'_0[z_n]$, $b \in I'$. This shows that $f, f_1, f_2, \ldots, f_n$ generate I.

I.C PROPOSITION The local ring O_0 of the origin in $\mathbb{C}^n$ is a unique factorization domain.

It suffices to prove the following: Suppose that f is an irreducible germ in O_0, g, h are in O_0, $f|gh$ and $f\nmid h$ then $f|g$. We may assume that f is a distinguished pseudo-polynomial in $O'_0[z_n]$, and that g, h are in $O'_0[z_n]$ as well, by the division theorem. By induction, and the lemma of Gauss, $O'_0[z_n]$ is factorial and the theorem follows from this.

I.D PROPOSITION (Nullstellensatz for hypersurfaces) Suppose that $X = V(f)$ is an analytic set in a polydisk about the origin in $\mathbb{C}^n$, defined by the one equation. Then the ideal of functions vanishing on X at the origin I_0 is rad (f).

Write $f = u f_1^{d_1} \cdots f_r^{d_r}$, u a unit, f_i irreducible. Then $V(f) = V(f_1) \cup \ldots \cup V(f_r)$. We want to prove that the ideal of functions vanishing on $V(f_i)$ is (f_i). Suppose that g vanishes on $V(f_i)$. We may assume that f_i is a distinguished pseudo-polynomial and write $g = af_i + b$, b a pseudo-polynomial of degree less than $\deg(f_i)$. If $b \neq 0$ then f_i and b must be coprime, so $\mathrm{Res}(f_i, b) = \varphi \in O'_0$ is non-zero and vanishes on $V(f_i)$, since both f_i and b do. Writing

$$f_i = z_n^p + a_{p-1}(z')\, z_n^{p-1} + \ldots + a_0(z')$$

one sees that there are points (z', z_n) in $V(f)$ with z' arbitrary, so long as it is small enough. But we can take $z' \notin V(\varphi) \subset C^{n-1}$, so we must have $b = 0$.

In this proof we have implicitly used the fact that the complement of a proper analytic subset of a polydisk is dense. It is also important to know that such a set is connected. For proofs of these facts, see Chapter I of Gunning - Rossi.

I. E. PROPOSITION Suppose $f, y \in O_0$ are coprime. Then they are coprime in a neighborhood of 0.

Again we assume that f is a distinguished pseudo-polynomial. Write $g = af + b$. We must prove that f, b are coprime in a neighborhood. Again form $\mathrm{Res}(f, b) = \varphi \in O'_0$. Let U be the neighborhood of 0 where φ is defined and in the ideal generated by f, b. If f and b have a common factor h not a unit at a point $a \in U$, then $h | \varphi$ which implies that V(f) contains, in a neighborhood of 0, a hypersurface of the form $h(z_1, \ldots, z_{n-1}) = 0$. But by writing

$$f = z_n^p + a_{p-1}(z')z_n^{p-1} + \ldots + a_0(z')$$

in U one sees that this cannot happen.

The geometric significance of the Weierstrass preparation theorem is this : Given an analytic hypersurface X through 0 in a polydisk about 0 in $\mathbb{C}^n$, one can choose a direction so that the projection of X from this direction onto a hyperplane is a finite analytic map from X onto a polydisk in $\mathbb{C}^{n-1}$, perhaps after restricting the first polydisk. A finite analytic map between analytic spaces is a proper, holomorphic map with finite fibers.

In this form the theorem generalizes: Given an analytic set X through 0 in a polydisk one can, by a series of projections, define a finite analytic map from X onto a polydisk in $\mathbb{C}^k$, some k less than n. (It may be necessary to shrink the polydisk).

This k depends only on X and the point 0, not the embedding into $\mathbb{C}^n$. It is called the dimension of X at 0. One can prove the following

I. F. THEOREM X is a complex manifold at 0 just in case there are functions $f_1, \ldots, f_k$, with $k = \dim_0 X$, which generate the maximal ideal of the local ring $O_{X,0}$.

More discussion of this is given in Gunning-Rossi [13]. The proof is essentially in the inverse function theorem.

From the existence of these branched coverings it is also possible to deduce that the set of non-singular points of X (that is, points at which X is a complex manifold) is open and dense. If U is the set of non-singular points of X then U breaks into components $U_1, \ldots, U_i, \ldots$. The closures $X_1 = \overline{U}_1$, $X_2 = \overline{U}_2, \ldots$ are analytic subvarieties of X, called the irreducible components of X. An analytic space is called irreducible if it has only one irreducible component. We shall almost always deal with irreducible analytic spaces. In particular, all our complex manifolds will be connected. More on these points appears in Narasimhan [26] .

We will discuss branched coverings again in Chapter III. For now, we'll give a small discussion of the local theory of algebraic varieties from a more algebraic point of view - just a statement of facts, running parallel to what we have seen analytically.

Just as a point on an analytic space has arbitrarily small neighborhoods which are analytic subvarieties of polydisks, a point on an algebraic variety has arbitrarily small (in the Zariski topology) Zariski neighborhoods which are closed subvarieties of $\mathbb{C}^n$ (such things are called affine varieties). An affine variety can be exhibited as a branched algebraic cover of $\mathbb{C}^k$, for some k, and the minimum of these k's as one runs over the neighborhoods of a point $x \in X$ is called the dimension of X at x. The definition of non-singularity on an algebraic variety is just like that given analytically, as is the definition of irreducibility. A word of warning may be necessary here: on an algebraic variety global irreducibility is equivalent to connectedness plus everywhere local irreducibility. This is not the case analytically, as one sees from the example

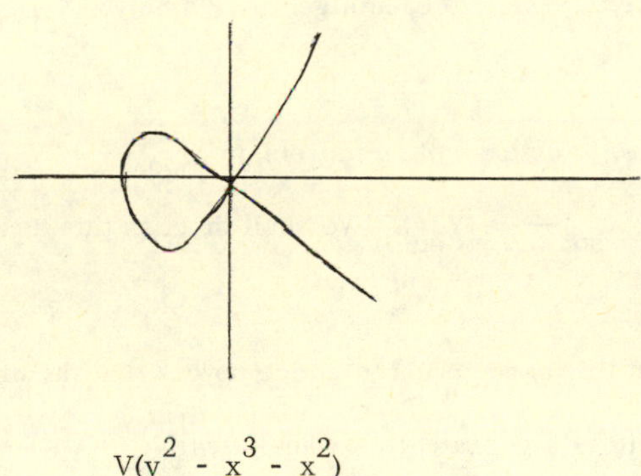

$$V(y^2 - x^3 - x^2)$$

which is globally irreducible yet reducible at the point $(0, 0)$.

Any algebraic subset X of $\mathbb{C}^n$ is also an analytic subset, and there is a natural morphism of ringed spaces

$$(X, O_{hol}) \longrightarrow (X, O_{alg}).$$

In case X is an arbitrary algebraic variety it is covered by affine opens, $X = \cup_i X_i$. For each i we get an analytic (X_i, O_{hol}) and a morphism

$$(X_i, O_{hol}) \longrightarrow (X_i, O_{alg})$$

Everything patches together to give a new topology on X, together with a sheaf of functions O_{hol} in this topology. This makes an analytic space, which will be Hausdorft in case the variety were separated. There will be a morphism of ringed spaces

$$(X, O_{hol}) \longrightarrow (X, O_{alg}).$$

The construction of (X, O_{hol}) can also be described intrinsically as follows: Let the new topology on X be the weakest such that all sections of O_{alg} are continuous as maps to $\mathbb{C}$ (in the strong topology on $\mathbb{C}$). Then let the sheaf O_{hol} have for sections over affines those functions which are uniform limits on compacts of sections of O_{alg}.

The association of an analytic space to each algebraic variety, $X_{alg} \longrightarrow X_{hol}$, has nice functorial properties. For example a morphism $X_{hol} \longrightarrow Y_{alg}$ defines in a natural way a morphism $X_{hol} \longrightarrow Y_{hol}$. We shall discuss this association in more detail in Chapter IV.

One important point about this association to notice now is that the dimension of X at a point is the same either algebraically or analytically.

Chapter One

§ 3 $\mathbb{P}^n$

Our first -and most important- examples of algebraic varieties which are not affine will be the projective spaces. The construction of projective spaces begins with the action of the multiplicative group $\mathbb{C}^x$ on the topological spaces $\mathbb{C}^{n+1} - \{0\}$ and $\mathbb{C}^{n+1}_{Zar} - \{0\}$. The quotients of these actions are denoted by $\mathbb{P}^n$ and $\mathbb{P}^n_{Zar}$. The maps

$$\mathbb{C}^{n+1} - \{0\} \xrightarrow{\varphi} \mathbb{P}^n \qquad\qquad \mathbb{C}^{n+1}_{Zar} - \{0\} \xrightarrow{\varphi} \mathbb{P}^n_{Zar}$$

are used to define sheaves of rings on $\mathbb{P}^n$, $\mathbb{P}^n_{Zar}$: For an open U in $\mathbb{P}^n$ or $\mathbb{P}^n_{Zar}$, $\Gamma(\varphi^{-1}(U), O_{hol})$ and $\Gamma(\varphi^{-1}(U), O_{alg})$ are acted upon by $\mathbb{C}^x$. Let $\Gamma(U, O_{\mathbb{P}^n hol})$ and $\Gamma(U, O_{\mathbb{P}^n alg})$ be the invariant elements of these rings. Thus

$$\Gamma(U, O_{\mathbb{P}^n hol}) = \{\text{holomorphic functions on } \varphi^{-1}(U) \text{ such that} \\ f(\lambda z) = f(z) \quad \text{for all } \lambda \in \mathbb{C}^x\}$$

and similarly in the algebraic case. This suffices to define ringed space structures on $\mathbb{P}^n$ and $\mathbb{P}^n_{Zar}$. Henceforth we shall not speak of $\mathbb{P}^n_{Zar}$ without its attendent sheaf to remind us of its topology, so we drop the subscript Zar.

The space $(\mathbb{P}^n, O_{\mathbb{P}^n hol})$ is an analytic space (in fact a complex manifold) and the space $(\mathbb{P}^n, O_{\mathbb{P}^n alg})$ is an algebraic variety. We shall verify the first of these state -ments. Note that in the map $\varphi: \mathbb{C}^{n+1} - \{0\} \longrightarrow \mathbb{P}^n$ the fiber over a point looks like $\{ (\lambda z_o, \ldots, \lambda z_n), \lambda \in \mathbb{C}^x \}$ for some $(z_o, \ldots, z_n) \in \mathbb{C}^{n+1} - \{0\}$. The n+1 - tuple

$(z_0, \dots, z_n)$ uniquely determines a point and is called a set of homogeneous coordinates for that point. Although one point in $\mathbb{P}^n$ has many sets of homogeneous coordinates, it is customary to speak of the point $(z_0, \dots, z_n)$. For $i = 0, 1, \dots, n$ the set of points $(z_0, \dots, z_n)$ with $z_i \neq 0$ is denoted $D^+(z_i)$. It is an open set and the $D^+(z_i)$ together form an open cover of $\mathbb{P}^n$. The map

$$D^+(z_0) \xrightarrow{\psi_0} \mathbb{C}^n$$

by
$$(z_0, \dots, z_n) \longrightarrow \left(\frac{z_1}{z_0}, \dots, \frac{z_n}{z_0}\right)$$

is a homeomorphism. For open $W \subset \mathbb{C}^n$ the $\mathbb{C}^x$ invariant elements of $\Gamma(\varphi^{-1} \circ \psi^{-1} \circ (W), O_{hol})$ are naturally identified with $\Gamma(W, O_{hol})$, so $D^+(z_0)$ is isomorphic as a ringed space to $\mathbb{C}^n$, and similarly for $D^+(z_i)$.

The analytic space associated to $(\mathbb{P}^n, O_{\mathbb{P}^n\,alg})$ is $(\mathbb{P}^n, O_{\mathbb{P}^n hol})$.

The analytic and algebraic subsets of $\mathbb{P}^n$ give further examples of analytic spaces and algebraic varieties: in analogy with what we did in $\mathbb{C}^n$, one defines analytic and algebraic subsets of analytic spaces and algebraic varieties and gives them a natural inherited ringed space structure.

The description of algebraic or analytic subsets of $\mathbb{P}^n$ if facilitated by the use of homogeneous coordinates. The map $\varphi : \mathbb{C}^{n+1} - \{0\} \longrightarrow \mathbb{P}^n$ is a morphism when $\mathbb{P}^n$, $\mathbb{C}^{n+1} - \{0\}$ are considered either as analytic spaces or algebraic varieties, so that for X an algebraic or analytic set in $\mathbb{P}^n$, $\varphi^{-1}(X)$ is an algebraic or analytic set in $\mathbb{C}^{n+1} - \{0\}$. In case X is algebraic one knows that $\varphi^{-1}(x) \cup \{0\}$ is an algebraic subset of $\mathbb{C}^{n+1}$, one which is invariant under the action of $\mathbb{C}^x$ on $\mathbb{C}^{n+1}$ and is thus defined by an homogeneous ideal in $\mathbb{C}[z_0, \ldots, z_n]$ - that is, an ideal generated by homogeneous elements. Also, any homogeneous ideal of the polynomial ring, which does not contain $(z_0, \ldots, z_n)$ in its radical, defines an algebraic subset of $\mathbb{P}^n$. As in the affine case, two homogeneous ideals define the same algebraic set in $\mathbb{P}^n$ just in case they have the same radicals. For more on these points, see Lefschetz [20], or Safarevic [35].

The algebraic set associated to the homogeneous ideal I will be denoted $V^+(I)$.

In the analytic case one can also show - although this is more difficult - that $\varphi^{-1}(x) \cup \{0\}$ is also an analytic subset of $\mathbb{C}^{n+1}$ (a theorem of Remmert and Stein). One can deduce from this and some other facts about analytic sets in $\mathbb{C}^{n+1}$ the theorem of Chow, that all analytic subsets of $\mathbb{P}^n$ are algebraic. We shall give some other proofs of this theorem later in these notes.

The simplest examples of algebraic (and analytic) subvarieties of $\mathbb{P}^n$ are those defined by single homogeneous equation $f(z_0, \ldots, z_n) = 0$. Consider, for example, the subset of $\mathbb{P}^2$ defined by

$$z_2^2 z_0^3 - z_1(z_1 - z_0)(z_1 - 2z_0)(z_1 - 3z_0)(z_1 - 4z_0) = 0$$

Model of the real projective plane, with algebraic subset described

Note that the intersection of this variety with $D^+(z_0)$ is the algebraic subset of $\mathbb{C}^2$ earlier pictured. As an analytic space, this equation defines a complex manifold which is a Riemann surface of genus two (for Riemann surfaces, see Gunning [14]).

As another example consider the algebraic subset of $\mathbb{P}^3$ given by $z_0^3 + z_1^3 + z_2^3 + z_3^3 = 0$. This will again be a complex manifold; an algebraic subset of this is given by $z_0^3 + z_1^3 + z_2^3 + z_3^3 = 0$, $z_0 z_1 - z_2 z_3 = 0$. This defines a Riemann surface of genus four. Finally, $z_0^3 + z_1^3 + z_2^3 = 0$ defines an analytic subset of $\mathbb{P}^3$ which is not a complex manifold.

Chapter Two

§ 1 Sheaves of modules

We shall deal extensively with sheaves of modules on a ringed space. Our treatment of basic definitions will be cursory and sometimes incomplete. For more details, see Godement [9].

What one means by an O - module on the ringed space (x, O) is a sheaf of abelian groups on x such that the sections over any open set are a module over the sections of O on that open, and such that the module structures are compatible with the restriction maps. We have already seen some examples of sheaves of modules in sheaves of ideals.

Some constructions dealing with sheaves of modules are:

If (X, O) is a ringed space with two sheaves of modules F and G one can form the sheaves $F \oplus G$, $\underline{\mathrm{Hom}}_O(F, G)$ and $F \otimes_O G$ by performing the indicated operation on sections over open sets to get presheaves, then taking the associated sheaves, which will have O-module structures.

Note that the global sections of $\underline{\mathrm{Hom}}_O(F, G)$ can be identified with the O-module homomorphisms from F to G.

Sheaves of modules can be pulled from one ringed space to another: If (X, O_X) and (Y, O_Y) are ringed spaces, $\varphi : X \longrightarrow Y$ a morphism and F a sheaf of modules on Y, there is a sheaf of modules $\varphi^* F$ on X, such that the stalk at a point $x \in X$, $(\varphi^* F)_x$, is

$F_{\varphi(x)} \underset{O_{Y,\varphi(x)}}{\otimes} O_{X,x}$. The operation φ^* is functorial. We shall be seeing a great deal of it later on.

Sheaf theory provides a way of dealing with questions on complex manifolds which involve local solutions and global patching. A basic feature of certain types of sheaves, which is of great importance inalgebraic geometry, is that in investi -gating local questions it is sometimes enough to work pointwise. This is the substance of a property called coherence.

A sheaf of modules M on a ringed space (X, O) is called coherent in case it has the two properties

(i) for any $x \in X$ there is an open neighborhood U of x over which there is a surjective map of O-modules

$$O^n|_U \longrightarrow M|_U$$

where O^n = direct sum of n copies of O. In other words, M is generated over O and U by a finite number of its sections over U.

(2) for any $x \in X$ and open neighborhood U of x over which there is a map of O-modules

$$O^n|_U \xrightarrow{\varphi} M|_U$$

there is a neighborhood W of x in U over which there is a surjective map of O - modules

$$O^m|_W \longrightarrow \mathrm{Ker}(\varphi)|_W$$

As an example of the use of the coherence condition, we prove the

II. A. THEOREM *Let M be coherent sheaf on a local ringed space* $(X,0)$. *Let* $x \in X$ *and suppose that* $m_1, \ldots, m_n \in M_x$ *generate* M_x *over* O_x. *Then there is some neighborhood U of x over which sections representing the* m_i *are all defined, and such that these sections generate the stalks at every point.*

Over an appropriate neighborhood W of x sections representing the m_i are defined and induce a map

$$O^n_{|W} \longrightarrow M_{|W}$$

which is surjective at the stalk at x. On a perhaps smaller neighborhood W' there is an exact sequence

$$O^i_{|W'} \longrightarrow O^{\ell}_{|W'} \longrightarrow M_{|W'} \longrightarrow 0$$

and a commutative diagram

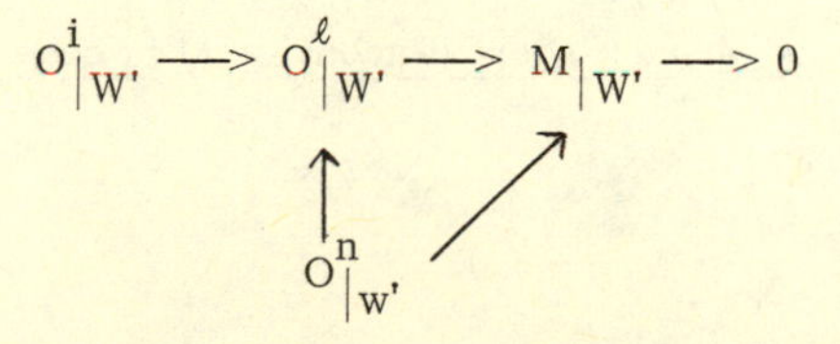

such that the map $O^{n+i}_{|W'} \longrightarrow O^{\ell}_{|W'}$ is surjective at x. This map is represented by a matrix of functions in O defined over W'. If we evaluate all functions at the point x we see that this matrix has an $\ell \times \ell$ invertible submatrix at the point x, and hence in a neighborhood since the ringed space is local (look at the determinant).

Then the map is surjective in a neighborhood of x.

II. B. COROLLARY <u>A coherent sheaf of modules on a local ringed space has closed support. The support is always given locally as the zeroes of distinguished functions.</u>

There is an important theorem which expresses the nice categorical properties of coherent sheaves.

II. C. THEOREM <u>Given an exact sequence</u>

$$0 \longrightarrow A \longrightarrow B \longrightarrow C \longrightarrow 0$$

<u>of</u> O-<u>modules on the ringed space</u> (X,O), <u>the coherence of any two of the sheaves implies the coherence of the third.</u>

Proofs appear in Serre [27] and Gunning-Rossi [13].

A ringed space on which the sheaf of rings is itself coherent is called an Oka ringed space.

II. D. PROPOSITION <u>On an Oka ringed space a sheaf of modules</u> M <u>is coherent just in case it is locally given as a cokernel</u>

$$O^n_{|U} \longrightarrow O^m_{|U} \longrightarrow M \longrightarrow 0$$

A proof appears in Serre[27].

A submodule of a coherent module is coherent just in case it is locally finitely generated (that is, satisfies condition 1 in the definition of coherence). Also, if a coherent module M on a ringed space is pulled back by a morphism $\varphi : X \longrightarrow Y$, then φ^*M is a coherent O_X- module. This follows from the right-exactness of tensoring. One applies this in particular to the case where $\varphi : X \longrightarrow Y$ is a closed immersion of analytic or algebraic varieties.

In the case of analytic spaces there is the important

II. E. THEOREM (Oka) A polydisk in $\mathbb{C}^n$ is an Oka ringed space.

We omit the proof of this. The basic analytic fact which is needed is the Weierstrass preparation theorem.

Another important theorem of Oka is

II. F. THEOREM The ideal sheaf of an analytic subset of a polycylinder is coherent.

From these last two theorems follows the

II. G. THEOREM An analytic space is an Oka ringed space.

All these theorems have analogues for algebraic varieties. The end result is the

II. H. THEOREM An algebraic variety is an Oka ringed space.

This theorem first appeared in Serre [27] It is much easier to prove than its analytic analogue. Proofs of the Oka theorems appear in Narasihman [26].

There are a few more important properties of coherent sheaves which we shall be using without proof. The reference again is Serre [27].

If F, G are coherent sheaves on a ringed space (X,O) then for $a \in X$,

$(\underline{\mathrm{Hom}}_0(F,G))_a \xrightarrow{\sim} \mathrm{Hom}_{0_a}(F_a, G_a)$ where the map is the one naturally defined.

Suppose that X is a closed analytic subvariety of an analytic variety Y. A sheaf F of O_X- modules is coherent as a sheaf of O_X-modules just in case it is coherent as a sheaf of O_Y-modules. An analogous statement is true in the algebraic case.

To illustrate some of these ideas, we give a discussion of singularities on algebraic and analytic varieties. If X is an analytic or algebraic variety, the diagonal map $X \xrightarrow{\Delta} X \times X$ is a closed imbedding; if I is the sheaf of ideals defining the imbedded X, we let $\Omega_{X/\mathbb{C}} = \Delta^*(I)$, and call this coherent sheaf the sheaf of <u>Kähler differentials</u> on X.

II. I. PROPOSITION <u>If</u> $a \in X$, <u>then</u> $(\Omega_{X/\mathbb{C}})_a / m_a(\Omega_{X/\mathbb{C}})_a$ <u>is isomorphic to</u> m_a/m_a^2 <u>as a</u> $\mathbb{C}$ - <u>vector space</u>, <u>where</u> m_a = <u>maximal ideal of</u> O_a.

This reduces to showing that if O is a local $\mathbb{C}$-algebra, with residue field $\mathbb{C}$ and maximal ideal m, then

$$m/m^2 \xrightarrow{\sim} I/I^2 \otimes_O \mathbb{C}.$$

<u>The map is given by</u>

$$g \longrightarrow 1 \otimes g - g \otimes 1$$

The proof that this is an isomorphism is purely algebraic and we omit it. See Mumford [25].

A word about the geometric meaning of the sheaf $\Omega_{X/\mathbb{C}}$ is in order: For a point a on an analytic or algebraic variety x, the vector space $(\Omega_{X/\mathbb{C}})_a/ma(\Omega_{X/\mathbb{C}})_a$ is to be thought of as the dual to the vector space generated by the tangent directions to X at a. We suppose that X is an analytic subvariety of an open set U in $\mathbb{C}^n$, and that a is the origin. A complex line through the origin is tangent to X at 0 in case it is a limit of secants through pairs of points on X as those pairs approach 0. An element f of $(\Omega_{X/\mathbb{C}})_a/m_a(\Omega_{X/\mathbb{C}})_a$ should define an hyperplane in the vector space generated by those tangent lines — we'll give a rough description of how this works.

A tangent line is defined by a sequence $\{(a_n, a_n')\}$ in $X \times X - \Delta$ approaching (0, 0) with n, and defining a line ℓ in the limit. ℓ is in the hyperplane defined by $f \in I_{(a,a)}/I^2_{(a,a)}$ ($I_{(a,a)}$ = stalk at (a, a) of diagonal ideal on $X \times X$) just in case

$$\lim_{n \to \infty} \frac{f(a_n, a_n')}{||(a_n, a_n')||} = 0.$$

Now we can investigate the geometric signifigance of the coherence of the sheaf $\Omega_{X/\mathbb{C}}$. First note that, for a coherent sheaf F on an analytic space or algebraic variety X, $\dim_{\mathbb{C}} F_a/m_a F_a$ is the minimal number of elements which generate F_a over O_a — this is a consequence of Nakayama's lemma. From coherence conditions one sees that $\dim_{\mathbb{C}} F_a/m_a F_a$ is an upper semi-continuous function of $a \in X$, in fact for any n the set $\{a: \dim_{\mathbb{C}} F_a/m_a F_a\} \geq n$ is defined locally by distinguished functions.

Consider the case of X an open set in $\mathbb{C}^n$. Here $\dim_{\mathbb{C}} (\Omega_{X/\mathbb{C}})_a/(\Omega_{X/\mathbb{C}})_a m_a$ = $\dim_{\mathbb{C}} m_a/m_a^2$ = n at every point.

One knows in general that for any local ring of a point O_a, $\dim_{\mathbb{C}} m_a/m_a^2$ = minimum number of generators of m_a over O_a (from Nakayama's lemma again).

From our earlier discussion of non-singularity we get the

II. J. THEOREM Let X be an analytic space. The set of points at which X is a complex manifold is the complement of an analytic subvariety.

From the definition of dimension one sees that $\dim_n X$ is upper semicontinuous in a, and that X is non-singular at a just in case $\dim_a x = \dim_{\mathbb{C}} (\Omega_{X/\mathbb{C}})_a / m_a (\Omega_{X/\mathbb{C}})_a$. If is irreducible then $\dim_a X$ is constant, say at n, since the regular locus is connected. Then the set of singular points is the set

$$\{a : \dim_{\mathbb{C}} (\Omega_{X/\mathbb{C}})_a / m_a (\Omega_{X/\mathbb{C}})_n\} \geq n.$$

The general case follows from this.

Similarly, there is the

II. K. THEOREM The set of points at which an algebraic variety is singular is an algebraic subvariety.

All this gives a nice interpretation of the behavior of the coherent sheaf $\Omega_{X/\mathbb{C}}$. Consider for example the analytic subset of $\mathbb{C}^2$ given by $y^2 = x^3 + x^2$

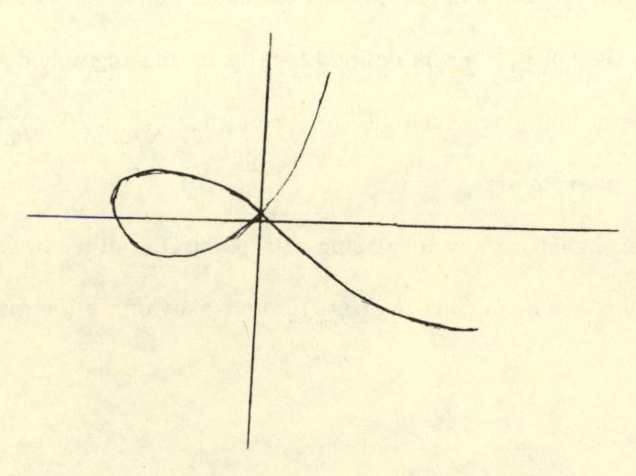

This will be a complex manifold everywhere except at (0, 0). $\dim_C m_a/m_a^2 = 1$, except for a = (0, 0) where the number is 2. Note that the tangent space to the variety at (0, 0) contains two linearly independent lines.

We'll now give a brief discussion of an aspect of the theory of analytic space (and algebraic varieties), the problem of the resolution of singularities. We state the problem formally:

Suppose X is an analytic space. Does there exist a complex manifold X' and a holomorphic map φ: X' —> X such that

(1) φ is proper and surjective

(2) If y is the singular locus of x, then φ: X - φ^{-1}(Y) —> X-Y is an isomorphism of complex manifolds.

This has been proven by Hironaka [15].

The simplest examples of the resolution of singularities is in the case of one-dimensional algebraic varieties. For example if x is the variety in $\mathbb{C}^2$ $V(z_2^2 - z_1^3)$ with singular point (0,0) then X = $\mathbb{C}$

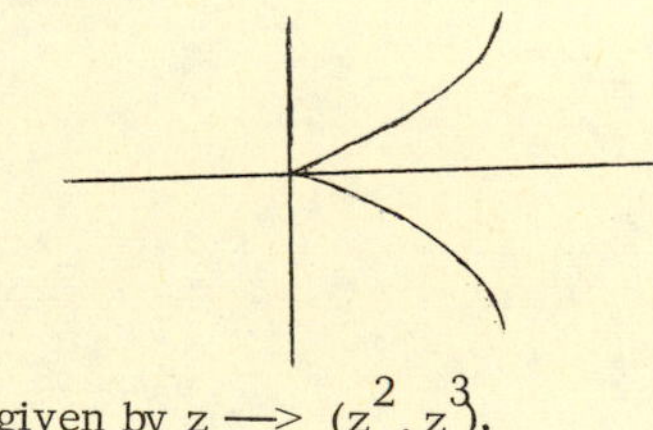

and the map $\mathbb{C}$ —> is given by z —> (z^2, z^3).

One needs more machinery than we have at our disposal to describe the interest -ing examples of the resolution of singularities in higher dimensions. We will mention one example of a surface: Let $X = V^+(X_1^2 + X_2^2 + X_3^2) \longrightarrow \mathbb{P}^3$, with a singularity at the point (1, 0, 0, 0). This singularity is resolved by an X, a complex manifold which is topologically $\mathbb{P}^1 \times \mathbb{P}^1$ but with a different complex structure.

Chapter Two

§ 2 Vector Bundles

An important class of coherent sheaves on analytic and algebraic varieties arises from vector bundles. We first recall the definition: If X is a topological space a (complex) vector bundle of rank n over X is given by a topological space over X, $\varphi : Y \longrightarrow X$, with the property that there is a covering $\{U_i\}$ of X such that there are isomorphisms

$$\varphi^{-1}(U_i) \xrightarrow[\eta_i]{\simeq} U_i \times \mathbb{C}^n$$

with φ restricting to the natural projection; it is further required that the maps

$$\eta_i \eta_j^{-1} : U_i \cap U_j \times \mathbb{C}^n \longrightarrow U_i \cap U_j \times \mathbb{C}^n$$

be linear on each fiber, so that there are continuous maps

$$\nu_{ij} : U_i \cap U_j \longrightarrow GL(n, \mathbb{C})$$

such that $\eta_i \eta_j^{-1} : (x, v) \longrightarrow (x, \nu_{ij}(x)v)$

If X is a differentiable manifold one gets the notion of a differentiable vector bundle by requiring all maps to be differentiable, using the natural differentiable structures on $\mathbb{C}^n$, $GL(n, \mathbb{C})$. If X is an analytic space one defines a holomorphic vector bundle by requiring Y to be an analytic space and all maps to be holomorphic (using the natural structures on $\mathbb{C}^n, GL(n, \mathbb{C})$).

A slight modification is required to define algebraic vector bundles over algebraic varieties: Here one requires Y to be an algebraic variety φ to be al gebraic. Also

$$\varphi^{-1}(U_i) \xrightarrow{\simeq}{\eta_i} U_i \times \mathbb{C}^n_{Zar}$$

with η_i algebraic. $GL(n, \mathbb{C})$ is a Zariski open of $\mathbb{C}^{n^2}$ and so has a natural algebraic structure; one requires the maps ν_{ij} to be algebraic.

If $Y \xrightarrow{\varphi} X$ is a vector bundle (continuous, differentiable, holomorphic or algebraic) one defines a sheaf of O_{cont}, O_{diff}, O_{hol}, or O_{alg} - modules $\Gamma(Y)$

$$\Gamma(U,\Gamma(Y)) = \text{sections of } \varphi : \varphi^{-1}(U) \longrightarrow U$$

These are sheaves of modules because one has the isomorphisms

$$\varphi^{-1}(U_i) \xrightarrow{\simeq} U_i \times \mathbb{C}^n \qquad (\text{or } \mathbb{C}^n_{Zar})$$

with which to multiply by functions pointwise and add; the linearity of the patching shows that this is unambiguously defined.

The sheaf $\Gamma(Y)$ has an important local triviality property. From

$$\varphi^{-1}(U_i) \xrightarrow{\simeq} U_i \times \mathbb{C}^n \quad (\text{or } \mathbb{C}^n_{Zar})$$

we see that $\Gamma(Y)\big|_{U_i}$ is locally free. In particular, $\Gamma(Y)$ is coherent if Y is a holomorphic on algebraic vector bundle. Thus in any of our four types of structure on X we have an association

$$\text{Vect}^n(X) \longrightarrow \{\text{isomorphism classes of locally free sheaves of rank n on X}\}$$

We want to show that this is an equivalence, that every locally free sheaf (up to isomorphism) arises in this way and that a vector bundle is determined (up to isomorphism) by its locally free sheaf.

We first take another look at the data which describes a vector bundle. Among other things we get an open covering $\{U_i\}$ and maps (continuous, differentiable, holomorphic, or algebraic).

$$\nu_{ij} : U_i \cap U_j \longrightarrow GL(n, \mathbb{C}).$$

Because of the relation between ν_{ij} and $\eta_i \eta_j^{-1}$ on $U_i \cap U_j$ we know that $\nu_{ik} = \nu_{ij}\nu_{jk}$ on $U_i \cap U_j \cap U_k$. Now suppose we are given a covering and maps ν_{ij}: $U_i \cap U_j \longrightarrow GL(n, \mathbb{C})$, satisfying $\nu_{ik} = \nu_{ij}\nu_{jk}$. Then we can construct a vector bundle by pasting together $U_i \times \mathbb{C}^n$ (or $U_i \times \mathbb{C}^n_{Zar}$) along the sets $U_i \cap U_j \times \mathbb{C}^n$ (or ________) in the obvious way; the condition on triple overlaps allows us to do this consistently. We call the set of all such maps $\{\nu_{ij}\}$ with respect to the covering $\{U_i\}$

$$Z^1(\{U_i\}, \ GL(n, O))$$

where O might be O_{cont}, O_{diff}, O_{hol}, or O_{alg}. We have shown that to every vector bundle which is defined according to the covering $\{U_i\}$ we can associate an element of Z^1, and conversely

A morphism of vector bundles

$$\varphi : Y \longrightarrow X, \quad \varphi' : Y' \longrightarrow X$$

is given by a map $\psi : Y \longrightarrow Y'$ such that

$$\begin{array}{ccc} \psi: \ Y & \longrightarrow & Y' \\ & \searrow \ \swarrow & \\ & X & \end{array}$$

commutes, and with the property that there is a covering $\{U_i\}$ of X according to which both Y and Y' are defined and such that the map

$$U_i \times \mathbb{C}^n \longrightarrow U_i \times \mathbb{C}^n$$

induced by the trivializations η_i, η_i' is of the form

$$(x, v) \longrightarrow (x, \varphi_i(v))$$

where φ_i: $U_i \longrightarrow \mathbb{C}^{n^2} = Mat_{nxn}(\mathbb{C})$. Note that the maps φ_i must satisfy

$$\nu_{ij} \varphi_j = \varphi_i \nu'_{ij} \qquad \text{on } U_i \cap U_j.$$

(All maps will be required to be differentiable, continuous, holomorphic, or algebraic, according to context.)

Conversely, from a collection of maps $\{\varphi_i\}$ satisfying

$$\nu_{ij}\varphi_j = \varphi_i \nu'_{ij}$$

we can construct a morphism of vector bundles.

From this we see that a necessary and sufficient condition for two elements $\{\nu_{ij}\}$ and $\{\nu'_{ij}\}$ of Z to define isomorphic vector bundles is the existence of maps $\varphi_i : U_i \longrightarrow GL(n, C)$ such that

$$\nu_{ij}\varphi_j = \varphi_i \nu'_{ij}$$

We say that two elements of $Z^1(\{U_i\}, GL(n, O))$ are equivalent just in case such maps exist; the quotient by this relation is called $H^1(\{U_i\}, GL(n, O))$.

Suppose that $\{W_\ell\}$ is a refinement of $\{U_i\}$, so that for each ℓ there is $\rho(\ell)$ such that $W_\ell \subset U_{\rho(\ell)}$. Then there is defined a map

$$H^1(\{U_i\}, GL(n, 0)) \longrightarrow H^1(\{W_\ell\}, GL(n, 0))$$

The direct limit of these sets over all coverings is denoted $H^1(X, GL(n, 0))$. It is equivalent to the set of isomorphism classes of (continuous, differentiable, holomorphic, or algebraic) vector bundles on X.

Now we will show how, given a locally free sheaf of constant rank, to associate an element of $H^1(X, GL(n, 0))$ to it. If L is that sheaf (of rank n) pick an open cover $\{U_i\}$ so that there are isomorphisms of O-modules restricted to U_i

$$L|_{U_i} \xrightarrow[\sim]{\eta_i} O^n|_{U_i}$$

On $U_i \cap U_j$ define ν_{ij} by the commutative diagram

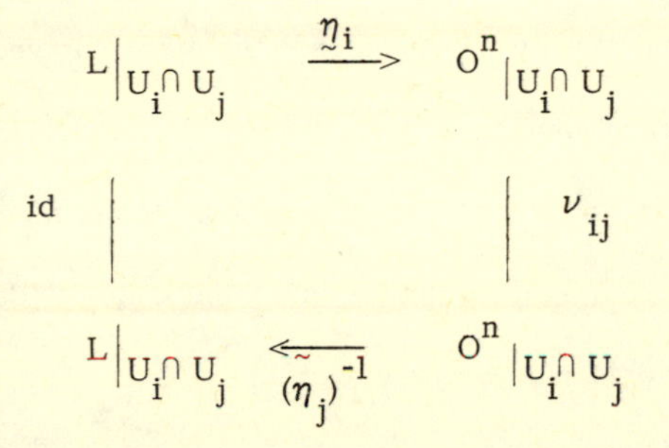

Then ν_{ij} is nothing but a map

$$\nu_{ij} : U_i \cap U_j \longrightarrow GL(n, \mathbb{C})$$

of the appropriate type. And $\nu_{ik} = \nu_{ij} \nu_{jk}$ on $U_i \cap U_j \cap U_k$, defining an element of $H^1(X, GL(n,O))$. The same sort of analysis as we just went through with vector bundles shows that $H^1(X, GL(n,O))$ is equivalent to {isomorphism classes of locally free sheaves of rank n}.

This is consistent with the previous association from vector bundles to locally free sheaves - so these notions are essentially equivalent.

This correspondence gives an easy way to define algebraic operations on vector bundles: Given bundles F, G to form $F \otimes G$ or <u>Hom</u> (F, G) we form the locally free sheaves $\Gamma(F) \otimes_O \Gamma(G)$, etc. , then take the associated bundles. This also allows us to define pull-backs of vector bundles with respect to maps $f: X \longrightarrow Y$.

We'll give some examples of vector bundles. One of the more important is the tangent bundle on a complex manifold. Given a complex manifold X we have a covering $\{U_i\}$ and holomorphic maps $\varphi_i: U_i \longrightarrow \mathbb{C}^n$ which are biholomorphic isomorphisms onto open sets in $\mathbb{C}^n$. Define ν_{ij} on $U_i \cap U_j$ as the Jacobian matrix of

$$\varphi_i \circ \varphi_j^{-1} : \varphi_j(U_j) \cap \varphi_i(U_i) \longrightarrow \varphi_j(U_j) \cap \varphi_i(U_i)$$

This set of maps satisfies the compatibility condition, so a holomorphic vector bundle (called the holomorphic tangent bundle) is defined on X, denoted T(X). As the name suggests, it is independent of the covering chosen. It has nice functorial properties: If $f : X \longrightarrow Y$ is a morphism of complex manifolds then there is a natural morphism of holomorphic bundles over X,

$$T(X) \longrightarrow f^* T(Y)$$

We'll just show how to define this locally. If X is an open set in $\mathbb{C}^n$, Y an open set in $\mathbb{C}^m$, then T(X), T(Y), and $f^*T(Y)$ are trivial bundles (that is, isomorphic to $X \times \mathbb{C}^n$, etc.). The map f is given by m holomorphic functions $f_1, \ldots, f_m$; the map of trivial bundles is given by the matrix $\left(\frac{\partial f_i}{\partial z_j}\right)$.

From consideration of the tangent bundle one derives another important bundle, the normal bundle of an imbedded submanifold: Given a closed imbedding of a complex manifold $f: Y \longrightarrow X$ one knows from local considerations that T(Y) is locally a direct summand of $f^*T(X)$; then there is a quotient bundle on Y,

$$f^*T(X)/T(Y) = N_{X/Y}$$

the normal bundle of Y in X. The rank of the normal bundle will be dim X -dim Y = co dim Y.

To appreciate the geometric significance of the normal bundle, note that if $M \longrightarrow Y$ is a holomorphic vector bundle (of rank n, say) then M naturally contains Y as a closed submanifold : Y is imbedded as the zero section of M. And the normal bundle of Y in M is just M.

It is a theorem in differential topology (see Milnor [23]) that in case $Y \longrightarrow X$ has normal bundle M then there is a neighborhood of Y in X which is differentiably isomorphic to a neighborhood of Y in M. But it will be in general impossible to pick holomorphically equivalent neighborhoods. In any case, the normal bundle of an imbedding gives some information about a neighborhood of Y in X.

On a non-singular algebraic variety X the sheaf $\Omega_{X/\mathbb{C}}$ will be locally free, thus defining a vector bundle on that variety, the <u>bundle of differentials.</u> The dual of this bundle is called the tangent bundle on a non-singular algebraic vareity; to justify this definition we observe that on a complex manifold X $\Omega_{X/\mathbb{C}}$ is naturally isomorphic to the sheaf of sections of the dual of the holomorphic tangent bundle.

To see this, note that the normal bundle of $X \xrightarrow{\Delta} X x X$ is just T(X). Now if $Y \longrightarrow X$ is a closed imbedding of complex manifolds with I the sheaf of ideals defined by Y then I/I^2 is a coherent sheaf on Y. There is an isomorphism of O_Y-modules

$$\underline{\text{Hom}}\,(I/I^2, O_y) \xrightarrow{\sim} N_{X/Y}$$

The idea is to get a map $T_{X|Y} \longrightarrow \underline{\text{Hom}}\,(I/I^2, O_Y)$ by differentiating a function in I along a tangent direction. Of course if the direction is tangent to Y the derivative will be 0, so the kernel of this map is just T_Y.

Applying this to the diagonal ideal brings us back to our previous viewpoint on $\Omega_{X/\mathbb{C}}$.

An important special class of vector bundles is the class of line bundles, or vector bundles of rank one. The set classifying them is

$$H^1(X, GL(1,O)) = H^1(X, O^x)$$

Two locally free sheaves of rank one L_1, L_2 are multiplied by

$$(L_1, L_2) \longrightarrow L_1 \underset{O}{\otimes} L_2$$

This is also a locally free sheaf of rank one, so we have defined a multiplication of line bundles. The inverse of a locally free sheaf of rank one, L_1, is $\underline{Hom}_O(L_1, O)$ - because $\underline{Hom}_O(L_1, O) \otimes_O L_1 \xrightarrow{\sim} O$. Because of the existence of inverse, locally free sheaves of rank one are called invertible sheaves.

One geometric way in which line bundles arise is in the consideration of Cartier divisors: an effective Cartier divisor on an analytic or algebraic variety X is given by a covering $\{U_i\}$ of X and for each i an $f_i \in \Gamma(U_i, O)$ such that f_i is not a divisor of zero in any stalk of U_i, and so that $\nu_{ij} f_j = f_i$ on $U_i \cap U_j$, ν_{ij} a unit in $\Gamma(U_i \cap U_j, O)$. Two effective Cartier divisors $\{(U_i, f_i)\}$ and $\{W_j, g_j)\}$ are the same in case $\varphi_{ij} g_j = f_i$ on $U_i \cap W_j$, φ_{ij} a unit in $\Gamma(U_i \cap W_j, O)$. Now from an effective Cartier divisor $\{(U_i, f_i)\}$ we define an invertible sheaf with transition functions $\nu_{ij} = \frac{f_i}{f_j}$. This definition does not depend on the exact data used to define the divisor, only on the divisor itself.

Two effective Cartier divisors are said to be linearly equivalent in case they define isomorphic line bundles. For the geometric content of this terminology see Lefschetz [20] or Safarevic [35].

Suppose that L is an invertible sheaf on the analytic or algebraic variety X. Any global section of L, $\sigma \in \Gamma(X, L)$, which is locally not a divisor of 0, defines an effective Cartier divisor; two sections define the same divisor just in case they differ by multiplication by a global unit. Conversely, any effective Cartier divisor with invertible sheaf L defines such an element of $\Gamma(X,L)$.

To see this, suppose the line bundle is defined by $\{U_i\}$, $\{\nu_{ij}\}$. Then σ is defined by $\sigma_i \in \Gamma(U_i, O)$ with $\sigma_i = \nu_{ij} \sigma_j$. The σ_i are not divisors of zero, so this defines a Cartier divisor. It is obvious how an effective Cartier divisor $\{U_i, f_i)\}$ defines a section of its line bundle.

An effective Cartier divisor is a subvariety of X which is locally defined by a single equation - it defines a coherent sheaf of ideals which is locally free of rank one, by $I(U_i) = f_i\Gamma(U_i, O)$. We could define an effective Cartier divisor to be a sheaf of ideals which is locally free of rank one. Then the sheaf of ideals defined by an effective Cartier divisor is isomorphic (as an invertible sheaf) to $L^{-1} = \underline{Hom}_{\mathcal{O}}(L, \mathcal{O})$, where L is the invertible sheaf associated to the divisor.

We have not emphasized the distinction between the different ideals which define the same analytic or algebraic subvariety, but it is important to keep this in mind when dealing with Cartier divisors. For example, $D_1 = (\mathbb{C}^2, z_1)$ and $D_2 = (\mathbb{C}^2, z_1^2)$ define, from our point of view, the same subvariety of $\mathbb{C}^2$, but different Cartier divisors. We think of D_2 as the subvariety D_1 doubled.

To give more justification for this, we define another notion of divisor: If X is an algebraic variety the group of Weil divisors on X is the free abelian group generated by the subvarieties of codimension one. One can show that in the case of a non-singular variety there is a natural map

$$\{\text{effective C-divisors}\} \longrightarrow \{\text{W-divisors}\}$$

which is an injection, and in fact exhibits the group of W-divisors as generated by the image of this map.

In the case of an algebraic variety with singularities one may not even be able to define the map properly. One difficulty is that not necessarily every subvariety of codimension one is given locally by a single equation. For example, the point $a = (0, 0)$ on the variety $V(y^2-(x^3+x^2))$ pictured before is a subvariety of codimension one, but if the ideal of functions vanishing at this point could be generated by one element then one would also have $\dim_{\mathbb{C}} m_a/m_a^2 = 1$, while we have seen that this dimension is two.

We'll give some more concrete examples of line bundles now. The most important line bundles in algebraic geometry are certain line bundles on projective spaces, both algebraic and holomorphic. Pick homogeneous coordinates $(z_0, \ldots, z_n)$ in $\mathbb{P}^n$ and $(z_0, \ldots, z_{n+1})$ in $\mathbb{P}^{n+1}$. Let p denote the point $(0, 0, \ldots, 0, 1)$ in $\mathbb{P}^{n+1}$, and define a map from $\mathbb{P}^{n+1} - \{p\}$ to $\mathbb{P}^n$ by

$$\varphi : (z_0, \ldots, z_{n+1}) \longrightarrow (z_0, \ldots, z_n)$$

This map (called the projection from p) is both an algebraic and holomorphic map. The preimage of a point $(z_0, \ldots, z_n)$ is the set of all points $(z_0, \ldots, z_n, \lambda)$, $\lambda \in \mathbb{C}$ - in other words, a line. In fact, $\varphi^{-1}\{D^+(z_0)\} \xrightarrow{\sim} \mathbb{C}^{n+1}$ and the map to $D^+(z_0)$ corresponds to projection on the first n terms. Similarly, for all i, $\varphi^{-1}\{D^+(z_0)\}$ is naturally isomorphic to $D^+(z_0) \times \mathbb{C}$. On $D^+(z_i z_j)$ the transition function associated to these natural maps is $\theta_{ij} = \frac{z_j}{z_i}$. The line bundle thus defined is called the tautological line bundle on projective space - either algebraically or holomorphically. Its sheaf of sections is denoted either $O_{\mathbb{P}^n}(1)_{alg}$ or $O_{\mathbb{P}^n}(1)_{hol}$. As a matter of notation, the dual of this sheaf will be denoted $O_{\mathbb{P}^n}(-1)_{alg}$ or $O_{\mathbb{P}^n}(-1)_{hol}$; the m-th tensor power of $O_{\mathbb{P}^n}(1)$ or its dual will be denoted $O_{\mathbb{P}^n}(m)$ or $O_{\mathbb{P}^n}(-m)$, so $O_{\mathbb{P}^n}(m) \otimes O_{\mathbb{P}^n}(q) = O_{\mathbb{P}^n}(m+q)$ for all m, $q \in Z$, provided we understand $O_{\mathbb{P}^n}(0)$ = the structure sheaf.

There is an alternative description of the sheaves $O_{\mathbb{P}^n}(m)$, in both the algebraic and holomorphic cases. We have the maps $\psi : \mathbb{C}^{n+1} - \{0\} \longrightarrow \mathbb{P}^n$, $\psi : \mathbb{C}^{n+1}_{Zar} - \{0\} \longrightarrow \mathbb{P}^n$ which define the structures on $\mathbb{P}^n$. For $m \in \mathbb{Z}$ we define a sheaf of $O_{\mathbb{P}^n}$-modules F(m) -

either algebraically or holomorphically - by

$$\Gamma(U, F(n)) = \{\text{algebraic or holomorphic functions } f \text{ on } \psi^{-1}(U) \text{ such that } f(\lambda z) = \lambda^m f(z) \text{ for all } z \in \psi^{-1}(U)\}.$$

Now the sheaf $F(m)$ is locally free of rank one: On $D^+(z_0)$ the map $\psi: \psi^{-1}(D^+(z_0)) \longrightarrow D^+(z_0)$ reduces to $\mathbb{C}^{n+1} - V(z_0) \xrightarrow{\psi} \mathbb{C}^n$, by

$$(z_0, \dots, z_n) \longrightarrow (\frac{z_1}{z_0}, \dots, \frac{z_n}{z_0}).$$

Given U open in $\mathbb{C}^n$, f algebraic or holomorphic on U, define $\tilde{f}$ on $\psi^{-1}(U)$ by

$$\tilde{f}(z_0, \dots, z_n) = z_0^m f({}^{z_1}/_{z_0}, \dots, \frac{z_n}{z_0}); \text{ then } \tilde{f} \in \Gamma(U, F(m)) \text{ we define}$$

$f \in \Gamma(U, O_{\mathbb{P}^n})$ by

$$f(\frac{z_1}{z_0}, \dots, \frac{z_n}{z_0}) = \tilde{f}(1, \frac{z_1}{z_0}, \dots, \frac{z_n}{z_0})$$

This defines an isomorphism of sheaves on $D^+(z_0)$. The same thing happens over $D^+(z_i)$, so this sheaf is invertible. The transition functions θ_{ij} associated with $F(m)$ with respect to the covering $\{D^+(z_i)\}$ are $\theta_{ij} = \frac{z_j}{z_i}$ - so that $F(m) = O_{\mathbb{P}^n}(m)$. All this is either algebraic or holomorphic.

There is a natural homomorphism

$$\Gamma(\mathbb{P}^n, O_{\mathbb{P}^n}(m)_{alg}) \longrightarrow \Gamma(\mathbb{P}^n, O_{\mathbb{P}^n}(m)_{hol})$$

of $\mathbb{C}$-vector spaces, the definition of which is obvious from the last definition of these sheaves. We'll prove a little theorem comparing the analytic and algebraic situation

III. X THEOREM The map

$$\Gamma(\mathbb{P}^n, O_{\mathbb{P}^n}(m)_{alg}) \longrightarrow \Gamma(\mathbb{P}^n, O_{\mathbb{P}^n}(m)_{hol})$$

is an isomorphism. There is a natural isomorphism

$$\Gamma(\mathbb{P}^n, \mathcal{O}_{\mathbb{P}^n}(m)_{alg}) \xrightarrow{\sim} \{\text{homogeneous polynomials of degree } m \text{ in } z_0, \ldots, z_n\}$$

An element f of $\Gamma(\mathbb{P}^n, \mathcal{O}_{\mathbb{P}^n}(m)_{hol})$ is a holomorphic function on $\mathbb{C}^{n+1}-\{0\}$ such that $f(\lambda z) = \lambda^m f(z)$ for all z. By Hartogs' theorem f is holomorphic in $\mathbb{C}^{n+1}$. (This shows already that f=0 unless $m \geq 0$). Represent f as a power series

$$f = \sum_{i_0, \ldots, i_n} \alpha_{i_0 \ldots i_n} z_0^{i_0} \cdots z_n^{i_n}$$

We must have $\alpha_{i_0 \ldots i_n} \lambda^{i_0 + \ldots + i_n} = \lambda^m \alpha_{i_0 \ldots i_n}$ by the uniqueness of power series representations. This shows that f is a homogeneous polynomial of degree m.

Chapter Two

§ 3 Sheaf cohomology and computations on $\mathbb{P}^n$.

Much of our work in these notes will use the tool of sheaf cohomology theory, which we will now recall. Details are in Godement [9], and Swan [34].

If A is a sheaf of abelian groups on a topological space X, there are defined, for all $i \geq 0$, the cohomology groups $H^i(X, A)$. They can be introduced as follows: A sheaf of abelian groups is called flabby if a section of the sheaf over any open set can be extended to a section over the full space. For any sheaf of abelian groups A on X there is an exact sequence extending indefinitely to the right

$$0 \longrightarrow A \longrightarrow F_0 \longrightarrow F_1 \longrightarrow F_2 \longrightarrow \dots \longrightarrow F_n \longrightarrow \dots$$

where the F_i are flabby. We set

$$H^i(X, A) = \ker(\Gamma(X, F_i) \longrightarrow \Gamma(X, F_{i+1})) / \operatorname{Im}(\Gamma(X, F_{i-1}) \longrightarrow \Gamma(X, F_i))$$

for $i > 0$, and $H^0(X, A) = \ker(\Gamma(X, F^0) \longrightarrow \Gamma(X, F_1))$

It can be shown that these groups do not depend on the particular flabby resolution of A. In case A is a module over some sheaf of rings 0, then one can choose the F_i to be 0-modules, so that the groups $H^i(X, A)$ are $\Gamma(X, 0)$-modules.

Some properties of the cohomology groups are:

There is always a natural isomorphism $H^0(X, A) \xrightarrow{\sim} \Gamma(X, A)$.

Any morphism of sheaves $A \longrightarrow B$ induces a morphism of cohomology groups $H^i(X, A) \longrightarrow H^i(X, B)$ for all i.

Given an exact sequence $0 \longrightarrow A \longrightarrow B \longrightarrow C \longrightarrow 0$ of sheaves of abelian groups there are defined, for all $i \geq 0$, maps $\delta_i: H^i(X, C) \longrightarrow H^{i+1}(X, A)$, so that the long sequence

$$0 \longrightarrow H^0(X, A) \longrightarrow H^0(X, B) \longrightarrow H^0(X, C) \xrightarrow{\delta} H^1(X, A) \longrightarrow \dots$$

is exact. In case the sheaves are sheaves of modules over a sheaf of rings, all the maps are $\Gamma(X, 0)$-homomorphisms.

A consequence of this is the following: Suppose

$$0 \longrightarrow A \longrightarrow B_0 \longrightarrow B_1 \longrightarrow B_2 \longrightarrow \dots$$

is an exact sequence of sheaves of abelian groups, such that $H^i(X, B_j) = 0$ for all $i > 0$, all j. Then

$$H^i(X, A) \xrightarrow{\sim} \mathrm{Ker}\,(H^0(X, B_i) \longrightarrow H^0(X, B_{i+1}))/(\mathrm{Im}(H^0(X, B_{i-1}) \longrightarrow H^0(X, B_i))$$

This makes it important to find cohomologically trivial sheaves, that is, sheaves F such that $H^i(X, F) = 0$ for $i > 0$. Flabby sheaves are cohomologically trivial.

Other important examples of such cohomologically trivial sheaves are the soft sheaves: A sheaf of abelian groups A on a Hansdorff space is called soft if to any covering $\{U_i\}$ of X there is a family $\{\phi_i\}$ of endomorphisms of A such that $\varphi_i = 0$ off U_i and only a finite number of the maps $\phi_{i,x}$ are non-zero at any stalk, and $\Sigma\, \phi_{i,x} = \mathrm{id}$ at any stalk. A partition of unity argument shows that the sheaf 0_{diff} on any differ -entiable manifold is soft; this will be the most important soft sheaf for us.

Closely related to the cohomology groups of sheaves are the Cech cohomology groups: First, from an open cover $\{U_i\}$ of X we define the groups $H^i(\{U_i\}, A)$ as follows: Define $C^i_{alt}(\{U_i\}, A)$ as the set of all maps f which to each i+1-tuple

$(U_{j_0},\ldots,U_{j_i})$ of opens in $\{U_i\}$ assigns an element $f_{j_0\ldots j_i} \in \Gamma(U_{j_0} \cap \ldots \cap U_{j_i}, A)$, in such a way that the association is alternating - that is, $f_{j_0\ldots j_i} = -f_{\ell_0\ldots\ell_i}$ if $(j_0,\ldots,j_i)$ and $(\ell_0,\ldots,\ell_i)$ differ by a transposition. C^i_{alt} has a natural group structure. There is a group homomorphism

$$d_i : C^i_{alt}(\{U_i\}, A) \longrightarrow C^{i+1}_{alt}(\{U_i\}, A)$$

given by $(d_i f)_{j_0\ldots j_{i+1}} = \Sigma_{k=0}^{i+1} (-1)^k f_{j_0\ldots \hat{j}_k \ldots j_{i+1}} \Big|_{U_{j_0} \cap \ldots \cap U_{j_{i+1}}}$

and the group $H^i(\{U_\ell\}, A)$ is defined as $\mathrm{Ker}(d_i)/\mathrm{Im}(d_{i-1})$ (We remark that one could drop the alternating requirements and work with the groups $C^i(\{U_\ell\}, A)$, doing the same thing. The resulting groups $H^i(\{U_\ell\}, A)$ would be isomorphic).

Given a refinement $\{W_j\}$ of $\{U_i\}$, so that $W_j \subset U_{p(j)}$, we can define maps

$$C^i_{alt}(\{U_\ell\}, A) \longrightarrow C^i_{alt}(\{W_j\}, A)$$

By taking the direct limit over all coverings, one gets the Cech cohomology groups $\check{H}^i(X, A)$.

It is not always the case that $\check{H}^i(X, A) = H^i(X, A)$ — in fact the cohomology groups $\check{H}^i(X, -)$ fail to have the nice exact sequence properties of the groups $H^i(X, -)$. The relation between these two cohomology theories is as follows: For any covering $\{U_i\}$ of X there is a spectral sequence with $E_2^{0,q}$ term $H^q(\{U_i\}, A)$, converging to $H^q(X, A)$. From consideration of these sequences one deduces

(1) For all i there are maps $H^i(\{U_j\}, A) \longrightarrow H^i(X, A)$, leading to maps $\check{H}^i(X, A) \longrightarrow H^i(X, A)$. For i = 0 or 1 this is always an isomorphism.

(2) Suppose $\{U_j\}$ is a covering such that

$$H^i(U_{j_1} \cap \ldots \cap U_{j_q}, A) = 0 \text{ for } i > 0 \text{ and any } j_1, \ldots, j_q \text{ in the index set.}$$

Then $H^i(\{U_j\}, A) \xrightarrow{\simeq} H^i(X, A)$

(Theorem of Leray).

For more discussion of this important point, see Godement [9].

There is a similarity between the Čech cohomology groups and the sets $H^1(X, GL(n, 0))$ which classify vector bundles. We can interpret these sets as cohomology sets of sheaves of non-abelian groups — and $H^1(X, GL(1, 0)) = H^1(X, 0^x)$, the first cohomology of the sheaf of units. The group structure on $H^1(X, 0^x)$ is the same as that otherwise defined on the line bundles.

It is by reference to the theorem of Leray that one can, in some cases, compute sheaf cohomology groups. What is needed is a class of cohomologically trivial open sets to make a covering. These are known to exist, in the case of coherent analytic sheaves, by the

II. L THEOREM (Cartan's Theorem B)

Suppose that F is a coherent analytic sheaf on the analytic space X, where X is a closed analytic subvariety of a polydisk in C^n. Then $H^i(X, F) = 0$ for i > 0.

This theorem has its algebraic analogue.

II. M THEOREM (Cartan's Theorem B-algebraic version)

Suppose that F is a coherent algebraic sheaf on the algebraic variety X, where X is a closed subvariety of C^n. Then $H^i(X, F) = 0$ for i > 0.

Since any analytic space or algebraic variety has a covering by such things, these theorems - together with the appropriate facts about intersections - tell us that cohomology can be computed by the Čech method. These theorems are proved in Gunning-Rossi [13] and Serre [27] . We shall discuss them more fully in Chapter Eight.

It is also sometimes possible to compute the cohomology of coherent analytic sheaves through the use of soft resolutions. This technique works when X is a complex manifold.

On a complex manifold M there is the sheaf Ω^1_{diff} of differentiable, complex-valued one-forms, which is a sheaf of 0_{hol}-modules (this is not a coherent sheaf). This sheaf splits into a direct sum of 0_{hol}-modules, $\Omega^1_{diff} = \Omega^{1,0} \oplus \Omega^{0,1}$: If U is an open polydisk in C^n then

$$\Omega^{1,0} = \text{subsheaf generated by } dz_1, \ldots, dz_n \text{ over } 0_{diff}$$

$$\Omega^{0,1} = \text{subsheaf generated by } d\bar{z}_1, \ldots, d\bar{z}_n$$

Any holomorphic automorphism of U takes $\Omega^{1,0}$ into itself and $\Omega^{0,1}$ into itself. This shows that there is such a decomposition on a complex manifold.

We get a corresponding splitting of s-forms $\Omega^s_{diff} = \sum_{p+q=s} \Omega^{p,q}$ where $\Omega^{p,q} = \Lambda^p(\Omega^{1,0}) \otimes \Lambda^q(\Omega^{0,1})$.

The operator of exterior differentiation splits into $d = \partial + \bar{\partial}$, where $\bar{\partial} : \Omega^{p,q} \longrightarrow \Omega^{p,q+1}$, $\partial : \Omega^{p,q} \longrightarrow \Omega^{p+1,q}$. Again, this is clear on a polydisk, where it is invariant under holomorphic automorphism.

An important local theorem is

II. 0 THEOREM (Poincaré lemma) <u>The sequence of sheaves</u>

$$0 \longrightarrow C \xrightarrow{\text{inj}} 0_{\text{diff}} \xrightarrow{d} \Omega^1_{\text{diff}} \xrightarrow{d} \cdots \longrightarrow \Omega^{2n}_{\text{diff}} \longrightarrow o$$

<u>is exact, if</u> $n = \dim M$.

As a consequence of this we get de Rham's theorem:

$$H^s(M, C) \xrightarrow{\sim} \{\text{d-closed s-forms}\} \;/\; \{\text{d-exact s-forms}\}$$

We want to investigate the similar situation for the $\bar{\partial}$ -operator. The $\bar{\partial}$-cohomology groups of the complex manifold M are defined by

$$H^{p,q}(M) = \{\bar{\partial}\text{ -closed p, q forms}\} \;/\; \{\bar{\partial}\text{ -exact p, q forms}\}$$

If N is a complex manifold, a holomorphic map $f: M \longrightarrow N$ induces maps $f^*: H^{p,q}(N) \longrightarrow H^{p,q}(M)$ for all p, q. This is functorial.

These maps depend only on the holomorphic homotopy class of f: Letting $\Delta = \{z \epsilon C: |z| < 1\}$, suppose we have a p, q-form ψ on $M x \Delta$, such that $\bar{\partial}\psi = 0$, of positive degree. Then we get forms on M by $\psi_0 = \psi\big|_{Mx\{0\}}$, $\psi_{1-\epsilon} = \psi\big|_{Mx\{1-\epsilon\}}$.

We will show that $\psi_0 - \psi_{1-\epsilon}$ is $\bar{\partial}$ -exact.

Letting w be a local holomorphic coordinate on M, we can locally write

$$\psi(w, z) = \alpha(z, w) - \beta(z, w) \wedge d\bar{z}$$

where $\alpha(z, w)$ involves no $d\bar{z}$. Then $\bar{\partial}\psi = 0$ implies that $\frac{\partial d}{\partial \bar{z}}(z, w) = \bar{\partial}_w \beta(z, w)$ and $\bar{\partial}_w \alpha = 0$. Set

$$\bar\partial \eta(w) = \int_0^{1-\epsilon} \bar\partial_w \beta(w,z) d\bar z = \int_0^{1-\epsilon} \frac{\partial \alpha}{\partial \bar z} = \alpha(w, 1-\epsilon) - \alpha(w,0).$$

Since $\psi \big|_{z=\text{constant}}$ is $\alpha \big|_{z=\text{constant}}$, we get what we wanted.

Now if F: $M \times \Delta \longrightarrow N$ is a holomorphic map of complex manifolds, this shows that the maps on $\bar\partial$ -cohomology $H^{p,q}(N) \longrightarrow H^{p,q}(M)$ induced by $F\big|_{M\times\{0\}}$, $F\big|_{M\times\{1-\epsilon\}}$ are the same. This implies the

II. P THEOREM ($\bar\partial$ -Poincaré lemma)

If w is a $\bar\partial$ -closed form of positive degree on the polydisk U, then w is $\bar\partial$ -exact on any smaller polydisk in U.

Let W be a smaller polydisk contained in U, so $zW \subset U$ if $|z| < 1+\epsilon$, some small $\epsilon > 0$, $z \in \mathbb{C}$. The map

$$W \times \Delta_{1+\epsilon} \longrightarrow U$$

given by $w \longrightarrow zw$ exhibits a holomorphic homotopy between the injection and the map to a point. This proves the theorem.

We can also formulate this theorem as:

The sequence of sheaves

$$0 \longrightarrow \Omega^p \longrightarrow \Omega^{p,0} \xrightarrow{\bar\partial} \Omega^{p,1} \longrightarrow \cdots \xrightarrow{\bar\partial} \Omega^{p,n} \longrightarrow 0$$

is exact.

Now we'll compute some cohomology groups by the Čech method.

THEOREM II. Q

$H^q(\mathbb{P}^n, \mathcal{O}_{\mathbb{P}^n_{hol}}(m)) = 0$, unless q=0 and $m \geq 0$, or q=n and $m \leq -n-1$. In these cases,

$$H^0(\mathbb{P}^n, \mathcal{O}_{\mathbb{P}^n_{hol}}(m)) \xrightarrow{\sim} \text{space of polynomials of degree } m \text{ in } z_0, \ldots, z_n$$

$$H^n(\mathbb{P}, \mathcal{O}_{\mathbb{P}^n_{hol}}(m)) \xrightarrow{\sim} \underline{\text{dual of}}\ H^0(\mathbb{P}^n, \mathcal{O}_{\mathbb{P}^n_{hol}}(-n-1-m)).$$

The part about the H^0's has already been done. The proof of the rest will use the fact that $H^i(\mathbb{C}\times\ldots\times\mathbb{C}\times\mathbb{C}^x\times\ldots\times\mathbb{C}^x, \mathcal{O}_{hol})=0$ for $i > 0$, which will be discussed in Chapter Eight. We shall also use the fact that any holomorphic function on $\mathbb{C}\times\ldots\times\mathbb{C}\times\mathbb{C}^x\times\ldots\mathbb{C}^x$ can be represented by a Laurent series in several variables,

$$f = \Sigma_{i_1,\ldots,i_n} \alpha_{i_1,\ldots i_n} z_1^{i_1}\ldots z_n^{i_n}$$

This can be proven, as in the 1-variable case, from Cauchy's integral formula. (Note that the existence of Laurent series representations on $\mathbb{C}^x$ is equivalent to $H^1(\mathbb{P}^1, \mathcal{O}_{hol})=0$).

Fix homogeneous coordinates $x_0, x_1, \ldots, x_n$ in $\mathbb{P}^n$ and identity $\mathbb{C}^n$ with $D^+(x_0)$, with affine coordinates

$$\frac{x_1}{x_0} = z_1, \frac{x_2}{x_0} = z_2, \ldots, \frac{x_n}{x_0} = z_n.$$ We shall use the notation $\underbrace{\mathbb{C}^x\times\mathbb{C}^x\times\ldots\times\mathbb{C}}_{n \text{ factors}}$ for $V(z_1z_2)$, etc. A holomorphic function f on $D(z_1)$ can be written $f = \Sigma_{\alpha_1,\ldots,\alpha_n} a_{\alpha_1\ldots\alpha_n} z_1^{\alpha_1}\ldots z_n^{\alpha_n}$; the condition that f be the restriction of a holomorphic function defined on $D^+(x_i)$ is that $\alpha_1 + \ldots + \alpha_n \leq 0$ for all non-zero monomials in the power series, while only α_1's can be negative.

Denote by N the covering $\{D^+(x_0), \ldots, D^+(x_n)\}$. By the theorem of Leray,

$H^q(N, F) = H^q(\mathbb{P}^n, F)$

Now if $q > 0$ an element of $Z^q_{alt}(N, \mathcal{O}_{\mathbb{P}^n}(d)_{hol})$ is given by a family of sections $f_{i_0 i_1\ldots i_q}$,

$i_o < i_1 < \ldots < i_q$, in $\Gamma(D^+(x_{i_0} \ldots x_{i_q}), \mathcal{O}_{\mathbb{P}^n}(d)_{hol})$ satisfying the cocycle rule. This data can be expressed as functions $f_{i_0 \ldots i_q}$ holomorphic in $D(Z_{i_0} \ldots Z_{i_q})$ (use $Z_o = 1$) satisfying for $i_0 < \ldots < i_{q+1}$

$$\left(\frac{z_{i_1}}{z_{i_0}}\right)^d f_{i_1 \ldots i_{q+1}} + \Sigma_{\ell=1}^{q+1} (-1)^\ell f_{i_0 \ldots \hat{i}_\ell \ldots i_{q+1}} = 0 \text{ with the}$$

restriction that $f_{i_0 \ldots i_q}$ extends to $D^+(x_{i_0} \ldots x_{i_q})$. (What we've done is systematically trivialize a section $f_{i_0 \ldots i_q} \in \Gamma(D^+(x_{i_0} \ldots x_{i_q}), \mathcal{O}_{\mathbb{P}}(d))$ as a function over a subset of $D^+(x_{i_0})$).

We want to show that any such thing is a coboundary, proved that $d \geq -n$. We want to find, for any $i_0 < \ldots < i_{q-1}$, $f_{i_0 \ldots i_{q-1}}$ holomorphic in $D(z_{i_0} \ldots z_{i_{q-1}})$ such that

$$(*) \quad f_{i_0 \ldots i_q} = \frac{z_{i_0}}{z_{i_1}}^d f_{i_1 \ldots i_q} + \Sigma_{\ell=1}^q (-1)^\ell f_{i_0 i_1 \ldots i_\ell \ldots i_q}$$

It is plain that a choice of $f_{0 i_1 \ldots i_{q-1}}$ for every $0 < i_1 < \ldots < i_{q-1}$ would together with the equations (*), determine all the other $f_{i_0 \ldots i_{q-1}}$. We could choose $f_{0 i_1 \ldots i_{q-1}}$ arbitrarily and get such a solution, defining

$$f_{i_0 i_1 \ldots i_{q-1}} = z_{i_0}^{-d} (f_{0 i_0 \ldots i_{q-1}} + f_{0 i_1 \ldots i_{q-1}} + \ldots + (-1)^q f_{0 i_0 \ldots i_{q-2}})$$

and the formal relations would be satisfied. But there is nothing to guarantee that $f_{i_0 \ldots i_{q-1}}$ comes by restriction from a function defined in $D^+(x_{i_0} \ldots x_{i_{q-1}})$. What must be checked is that in the Laurent series for

$$f_{0 i_0 \ldots i_{q-1}} + f_{0 i_1 \ldots i_{q-1}} + \ldots + (-1)^q f_{0 i_0 \ldots i_{q-2}}$$

there are no monomials of weight greater than d, so we must choose $f_{0i_1\ldots i_{q-1}}$, etc., so as to get rid of such terms in $f_{0i_0\ldots i_{q-1}}$. Now if $q<n$ there is, we may suppose, some $i_q > i_{q-1}$.

Then the relation

$$z_{i_0}^d f_{i_0 \ldots i_q} + \Sigma_{\ell=0}^q (-1)^{\ell+1} f_{0i_0\ldots \hat{i}_\ell \ldots i_q} = 0$$

shows that there can be in the expansion of $f_{0i_0\ldots i_{q-1}}$ no monomial of weight greater than d in which each of $z_{i_0}, \ldots, z_{i_{q-1}}$ appears to a negative power. In case $d \geq -n$, a counting argument shows the same for the case q=n.

If there is a monomial of weight greater than d in the expansion of $f_{0i_0\ldots i_{q-1}}$ in which only z_{i_0} of $z_{i_0}, z_{i_1}, \ldots, z_{i_{q-1}}$ does not appear to a negative power, then this determines a coefficient in the expansion of $f_{0i_1\ldots i_{q-1}}$. In case two or more of $z_{i_0}, \ldots, z_{i_{q-1}}$ appear to non-negative powers we get a more complicated condition, involving coefficients of monomials in two or more of the $f_{0i_1\ldots i_{q-1}}, f_{0i_0\ldots i_{q-1}}$. Suffice it to say that such conditions can be met consistantly (this amounts to computing some Čech cohomology with constant coefficients).

We have now shown that $H^q(\mathbb{P}^n, 0_{\mathbb{P}^n_{hol}}(d)) = 0$, for $q > 0$, except for the cases q=n, $d < -n$. We omit the computation showing

$$\dim_{\mathbb{C}} H^1(\mathbb{P}^1, 0_{\mathbb{P}^1_{hol}}(-2+n)) = n+1.$$

The rest of the theorem will go by induction.

Recall that $0_{\mathbb{P}^n_{hol}}(-1)$ is isomorphic to the sheaf of ideals of functions vanishing on the hyperplane $V^+(X_0)$, and that $\mathcal{O}_{\mathbb{P}^n}(1)\big|_{V^+(X_0)}$ is $\mathcal{O}_{\mathbb{P}^{n-1}}(1)$ on this $\mathbb{P}^{n-1}$.

Thus we have an exact sequence of sheaves

$$0 \longrightarrow \mathcal{O}_{\mathbb{P}^n_{hol}}(d-1) \longrightarrow \mathcal{O}_{\mathbb{P}^n_{hol}}(d) \longrightarrow \mathcal{O}_{\mathbb{P}^{n-1}_{hol}}(d) \longrightarrow 0$$

From what we have proved so far we get an exact sequence, for $n > 1$

$$0 \longrightarrow H^{n-1}(\mathbb{P}^{n-1}, \mathcal{O}_{\mathbb{P}^n_{hol}}(d)) \longrightarrow H^n(\mathbb{P}^n, \mathcal{O}_{\mathbb{P}^n_{hol}}(d-1)) \longrightarrow H^n(\mathbb{P}^n, \mathcal{O}_{\mathbb{P}^n}(d)) \longrightarrow 0$$

The rest of the theorem follows by inducting first over n, then over n-d. (Note that we have not exhibited any canonical duality between H^0's and H^n's. In fact there is one; see Serre [27]).

This proof also works for the groups $H^q(\mathbb{P}^n, \mathcal{O}_{\mathbb{P}^n_{alg}}(d))$; one simply works with truncated power series.

II. R THEOREM <u>For all</u> q, n, d $H^q(\mathbb{P}^n, \mathcal{O}_{\mathbb{P}^q_{alg}}(d)) \xrightarrow{\approx} H^q(\mathbb{P}^n, \mathcal{O}_{\mathbb{P}^n_{hol}}(d))$.

Note that the isomorphisms are canonically defined, if one computes the groups off the standard covering.

Chapter Three

§1 Maximum principle and Schwarz lemma on analytic spaces

Some basic facts in complex function theory are entirely local in character and these go over directly to the situation of a complex manifold. Often revised versions will hold for analytic spaces. One might expect that the way to generalize would be to resolve singularities, which is always possible. But it is easier to use the technique already mentioned in Chapter One of exhibiting an analytic space as locally a branched covering of a polydisk in $\mathbb{C}^n$. We now state the basic fact formally:

III. A. THEOREM If x is a point on an analytic space X there is a neighborhood U of x and a map $\pi : U \longrightarrow \Delta^n(r)$ for some r and n which is holomorphic and finite. That is, there is an analytic subset D of $\Delta^n(r)$ such that $\pi : U - \pi^{-1}(D) \longrightarrow \Delta^n(r) - D$ is a map of constant rank and a topological cover. The degree of the cover is called the degree of the map. We can assume that $\pi^{-1}(\{0\}) = \{x\}$.

For a proof, see Narasimhan [26].

Now we'll give some applications of this theorem.

III. B. PROPOSITION (Maximum principle in several variables)

Let $f: \Delta^n(r) \longrightarrow \mathbb{C}$ be a holomorphic function. If $|f|$ is a local maximum at 0 then f is constant.

Given $z \epsilon C^n$, define $f_z(u)$ for small $u \epsilon C$ by $f_z(u) = f(uz)$. By the one variable maximum principle, f_z is constant. The proposition follows from this.

Resolution of singularities easily gives an extension of this to analytic spaces. We can also do this by branched coverings.

III. C. PROPOSITION (Maximum principle for analytic spaces)

Let $f: X \longrightarrow C$ be a holomorphic function on an analytic space, with $|f|$ a local maximum at $x \in X$. Then f is constant in a neighborhood of X.

We may assume that X is irreducible and that there is $\pi : X \longrightarrow \Delta^n(r)$ with $\pi^{-1}\{0\} = \{x\}$, a d-sheeted analytic cover. For $z \in \Delta^n(r) - D$, we set

$$\varphi_k(z) = f(x_1(z))^k + \ldots + f(x_d(z))^k$$

where $\{x_1(z), \ldots, X_d(z)\} = \pi^{-1}\{z\}$. We may assume that f is a maximum at x_0, and $f(x_0) = 1$. Since $\varphi_k(z)$ is bounded it can be extended to a holomorphic function on $\Delta^n(r)$ (see Gunning-Rossi [13], Chapter I). Now

$$\varphi_k(0) = \mathrm{Vol}(\Delta^n(r))^{-1} \int_{\Delta^n(r)} \varphi_k(z) d\lambda$$

Suppose that f is not constant. Then it follows from the maximum principle on complex manifolds, using the irreducibility of X, that $|f| < 1$ on $X - \pi^{-1}(D)$, and from this we see that $\varphi_k(z) \xrightarrow[k]{} 0$ when $z \notin D$. Since D has measure 0, this implies $\varphi_k(0) \xrightarrow[k]{} 0$, while it must be that $\varphi_k(0) = 1$ identically.

The Schwarz lemma also admits generalization.

II. C. PROPOSITION (Schwarz lemma in several variables)

Let f be a holomorphic function on $\Delta^n(r)$ with $|f| \leq M$ everywhere and with a zero of order h at the origin. Then $|\varphi(z)| \leq M\left(\frac{|z|}{r}\right)^n$.

Fix $z \in \Delta^n(r)$, $z \neq 0$, and let $\psi_z(t) = \frac{f(tz)}{t^h}$, for $|t| < \frac{r}{|z|}$. Now $|\psi_z^{\bullet}(t)| \leq M\left(\frac{|z|}{r}\right)^h$ everywhere, as one sees by letting $|t| \longrightarrow \frac{r}{|z|}$ and applying the maximum principle. In particular, $|\psi_z(1)| = |\varphi(Z)| \leq M\left(\frac{|z|}{r}\right)^h$.

The Schwarz lemma for analytic spaces is stated in terms of branched coverings.

II. E. PROPOSITION (Schwarz lemma for analytic spaces)

Let $\pi : X \longrightarrow \Delta^n(r)$ be an analytic space exhibited as a d-sheeted branched covering of the polydisk. Let f be a holomorphic function on X with $|f| \leq M$ and suppose that $f \in (z_1, \ldots, z_n)^h$ in O_x, $\{x\} = \pi^{-1}\{(0, \ldots, 0)\}$ Then

$$|f(x)| \leq M \left(\frac{|\pi(x)|}{r}\right)^h \quad \text{for all } x \in X.$$

Once again set

$$\varphi_k(z) = f(x_1(z))^k + \ldots + f(x_d(z))^k$$

on $\Delta^n(r)$-D and extend. Even at a point $z \in D$, we have $\varphi_k(z) = f(x_1(z))^k + \ldots + f(x_j(z))^k$ although some x_i may be repeated. Applying the Schwarz lemma to φ_k, we get

$$|\varphi_k(z)| \leq M^k d \left(\frac{|z|}{r}\right)^{hk}$$

$$|\varphi_k(z)|^{1/k} \leq d^{1/k} M \left(\frac{|z|}{r}\right)^h$$

The proposition now follows from the

III.F. LEMMA Let $a_1, \ldots, a_d$ be complex numbers. Then

$$\lim_{k \to \infty} \left(|a_1|^k + \ldots + |a_d|^k\right)^{1/k} = \sup_i |a_i|$$

We omit the proof of this.

As a final example of the use of the technique of branched coverings we shall generalize the theorem about the completeness of the space of holomorphic functions.

III.G. PROPOSITION Let $V = V(f)$ be a hypersurface in the polydisk $\Delta^n(r)$. A sequence of holomorphic functions converging uniformly on compact subsets of V has a holomorphic limit.

We may assume that the origin is a point of V and it will suffice to show that the limit is holomorphic at the origin. Perhaps restricting to a slightly smaller polydisk, we may assume that f is a distinguished pseudo-polynomial

$$z_n^d + \alpha_{d-1}(z')z_n^{d-1} + \ldots + \alpha_0(z') = p(z', z_n).$$

The projection to the z' hyperplane exhibits V as a d-sheated branched covering which is ramified over those points where the discriminant of the polynomial vanishes - call this D. Given g analytic on V define $\varphi(z', z_n)$ for points in the polydisk with $z' \notin D$ by

$$\varphi(z', z_n) = \Sigma_{i=1}^{d} \frac{p(z', z_n)}{z_n - x_i(z')} g(x_i(z')).$$

After perhaps restricting to a slightly smaller polydisk there will be a constant M - independent of g - such that $||\varphi|| \leq M \, ||g||$. This implies that φ can be extended to the entire polydisk, since we may assume g bounded. The function will satisfy

$$\varphi(z', z_n) = \frac{\partial}{\partial z_n} p(z', z_n) g(z', z_n)$$

for $(z', z_n) \epsilon V$.

Now given a convergent sequence $\{g_n\}$ we can in a neighborhood of the origin associate to it a sequence $\{\varphi_n\}$, also convergent, with holomorphic limit φ. The relation $\varphi = \frac{\partial}{\partial z_n} p \; g$, with $g = \lim_n g_n$, shows that g is holomorphic.

Two remarks about this proof are in order: First, we have generalized the completeness theorem to hypersurfaces. The technique of branched coverings, properly applied, will extend the theorem to other codimensions - see Narasimhan [26]. Second, the same proof works under the weaker hypothesis that the

g_n are defined, holomorphic, and bounded only on the regular points of V. The sheaf of such functions is called the sheaf of weakly holomorphic on V and appears in the process of normalization. For more details, see Narasimhan [26].

Chapter Three

§ 2 Siegel's Theorem

Now we'll put some of these results to work and prove our first comparison theorem, a theorem about meromorphic functions. We first give an algebraic description of these objects: If X is an analytic space on algebraic variety one forms a presheaf $\tilde{K}_X$ by

$$\Gamma(U, \tilde{K}_X) = \text{total quotient ring of } \Gamma(U, 0_X)$$

The associated presheaf is denoted by K_X. In case X is an irreducible analytic space or algebraic variety, $\Gamma(X, K_X)$ will be a field, the field of meromorphic or rational functions on X. A meromorphic or rational function is thus defined by a covering $\{U_i\}$ and functions $\varphi_i \psi_j = \varphi_j \psi_i$ on $U_i \cap U_j$.

If X is an irreducible algebraic variety then K_X is a sheaf of fields and in fact a constant sheaf. We denote this field by $K_{rat}(X)$. It is a finitely generated extension of $\mathbb{C}$ with transcendence degree equal to the dimension of X. For details on this, see Lefschetz [20] or Safarevic [35].

Also if X is an irreducible algebraic variety the natural analytic structure on X (in which X is irreducible, as will be proven in the next section) gives rise to the field of meromorphic functions $K_{mer}(X)$ and there is an injection of fields $K_{rat}(X) \longrightarrow K_{mer}(X)$.

We want to show that, in case X is a complex manifold, meromorphic functions on X can be interpreted as quotients of sections of line bundles. First, given a line bundle with transition functions $\{\alpha_{ij}\}$ and two sections $\{\psi_i\}$, $\{\varphi_i\}$ defined with respect to the cover $\{U_i\}$ we can form a meromorphic function $\{\varphi_i / \psi_i\}$. On a complex manifold this procedure can be reversed. For, given a meromorphic function f we can, by the theorems in Chapter Two, find a covering $\{U_i\}$ of X so that on U_i $f = \frac{\varphi_i}{\psi_i}$,

where φ_i and ψ_i have no common factors except units (we can do this pointwise and automatically get the result in a neighborhood). Then $\varphi_i \varphi_j^{-1} = \psi_i \psi_j^{-1}$ is a unit in $U_i \cap U_j$ and these form the transition functions for an invertible sheaf, of which $\{\varphi_i\}$, $\{\psi_i\}$ are global sections.

Now that meromorphic functions have been defined we can extend our discussion of Cartier divisors. If X is an algebraic variety or analytic space a Cartier divisor on X is defined by a covering $\{U_i\}$ and $\varphi_i \epsilon \Gamma(U_i, K_X)$ which is a unit in $\Gamma(U_i, K_X)$, such that on $U_i \cap U_j$ $\varphi_i/\varphi_j \in \Gamma(U_i \cap U_j, O^x)$. Two sets of such data $\{U_i, \varphi_i\}$ and $\{W_j, \psi_j\}$ define the same divisor if $\{U_i, W_j, \varphi_i, \psi_j\}$ also defines a Cartier divisor. Cartier divisors form a group with the multiplication

$$\{U_i, \varphi_i\} \times \{W_j, \psi_j\} \longrightarrow \{U_i \cap W_j, \varphi_i \psi_j\}.$$

Also any Cartier divisor defines the transition functions of a line bundle, and the map from Cartier divisors to isomorphism classes of line bundles is a group homomorphism. In sheaf-theoretic terms we form the exact sequence

$$0 \longrightarrow O_X^x \longrightarrow K_X^x \longrightarrow K_X^x / O_X^x \longrightarrow 0$$

leading to the exact cohomology sequence

$$0 \longrightarrow H^0(X, O_X^x) \longrightarrow H^0(X, K_X^x) \longrightarrow H^0(X, K_X^x / O_X^x) \xrightarrow{\delta} H^1(X, O_X^x) \longrightarrow$$

Now $H^0(X, K_X^x/O_X^x)$ is the group of Cartier divisors and the map δ associates a line bundle to each Cartier divisor. The kernel of the map δ is just those Cartier divisors which come from meromorphic or rational functions; thus given two sections of a line bundle they each define an effective Cartier divisor and their quotient defines a meromorphic or rational function.

If X is an irreducible algebraic variety then $H^1(X, K_X^x)=0$, the sheaf K_X^x being constant. This shows that every algebraic line bundle on an irreducible algebraic variety is associated to some divisor.

The notions of divisor and meromorphic function appear already in Riemann surface theory. On $\mathbb{P}^1$, for instance, the groups of algebraic and analytic divisors are the same: each can be identified with the free abelian group generated by the points of $\mathbb{P}^1$, so a divisor is a finite sum $\sum_{p \in \mathbb{P}^1} n_p P$. The field of meromorphic functions on $\mathbb{P}^1$ is the same as the field of rational functions, $\mathbb{C}(z)$, and a divisor is the divisor of a meromorphic function just in case $\sum_{p \in \mathbb{P}^1} n_p = 0$. All these facts have their generalizations to arbitrary compact Riemann surfaces, relating to the theory of algebraic curves.

Our first theorem will generalize the identity between rational and meromorphic functions on a compact Riemann surface. It was proved by Siegel about 1950.

III. G. THEOREM <u>The field of meromorphic functions on an irreducible compact analytic space is an algebraic function field-that is, a finitely generated extension, of $\mathbb{C}$. The transcendence degree of this field over $\mathbb{C}$ is not greater than the dimension of the space.</u>

We'll give the proof only in case X is a complex manifold. What is needed is an application of the Schwarz lemma. In the general case one uses the Schwarz lemma for analytic spaces.

III. H. LEMMA <u>Let L be a holomorphic line bundle on the compact, irreducible complex manifold X. There is a finite covering $\{U_i\}$ of X such that</u>

(1) <u>any open</u> U_i <u>is holomorphically equivalent to a polydisk with center</u> $a_i \epsilon U_i$, say to $\Delta^n(\frac{3}{2})(a_i)$.
<u>Also the polydisks</u> $\Delta^n(e^{-1})(a_i)$ <u>cover</u> X.

(2) L <u>is defined according to this covering, with transition functions</u> α_{ij}

(3) <u>If we set</u>

$$\beta = \sup_{i, j \;\; \Delta^n(1)(a_i) \cap \Delta^n(1)(a_j)} \log |\alpha_{ij}|$$

<u>and</u>

h(L) = smallest integer greater than β

<u>then any global section of</u> L <u>which vanishes to order</u> h(L) <u>at each point</u> a_i <u>must be identically zero.</u>

First find a finite cover $\{U_i\}$ of X such that each U_i is holomorphically equivalent to the polydisk $\Delta^n(\frac{3}{2})$ with center a_i, such that L is defined according to this cover and X is covered by the smaller polydisks $\Delta^n(e^{-1})(a_i)$.

Suppose that there is a section $\{\varphi_i\}$ of L over X vanishing to order at least h(L) at each a_i. Let $M = \sup_{i, \; \Delta^n(1)(a_i)} |\varphi_i|$.

M is finite and reached at some point x, say $M = |\varphi_i(x)|$. Now $x \epsilon \Delta^n(e^{-1})(a_j)$ for some j. By the Schwarz lemma,

$$|\varphi_j(x)| \leq Me^{-h(L)}$$

But $|\varphi_i(x)| = |\alpha_{ij}(x)| \; |\varphi_j(x)|$ so we get

$$M \leq Me^{\beta - h(L)}$$

and this implies that M = 0 so $\{\varphi_i\} = 0$.

This lemma gives an upper bound for $\dim_{\mathbb{C}} H^0(X, L)$. For the condition that a global section of L vanish to order h at a point a_i defines a subspace of codimension less than or equal to

$$\Sigma_{\ell=0}^{h-1}\binom{n+\ell-1}{n-1} = \binom{n+h-1}{n} \leq h^n$$

this being the dimension of O_o^n / m_o^n. If there are m points a_i the lemma shows that

$$\dim_{\mathbb{C}} H^o(X, L) \leq m\ h(L)^n$$

and this is what we need to prove the theorem.

Given meromorphic functions $f_0, f_1, \ldots, f_n$ each is a quotient of two sections of a line bundle,

$$f_i = \frac{\varphi_i}{\psi_i}, \varphi_i, \psi_i \in H^o(X, L).$$

A polynomial $p(X_0, \ldots, X_n)$ of degree s_i in X_i defines a section of $L_o^{s_o} \otimes L_1^{s_1} \otimes \ldots \otimes L_n^{s_n}$, which will be the numerator of the meromorphic function $p(f_0, \ldots, f_n)$ by the rule

$$\Sigma\, \alpha_{i_o \ldots i_n} X_o^{i_o} \ldots X_n^{i_n} \longrightarrow \Sigma\, \alpha_{i_o \ldots i_n} \varphi_o^{i_o} \otimes \psi_o^{s_o - i_o} \otimes \ldots \otimes \varphi_n^{i_n} \otimes \psi_n^{s_n - i_n}$$

The meromorphic function $P(f_0, \ldots, f_n)$ will be zero just in case this section is zero.

We make use of a fact, that the covering $\{U_i\}$ can be chosen independently of the line bundle; this is a consequence of the fact (to be proven in Chapter Eight) that any line bundle on a polydisk is trivial. Now

$$h(L_o^s \otimes L_1^t \otimes \ldots \otimes L_n^t) \leq sh(L_o) + t\,h(L_1) + \ldots + th(L_n)$$

so

$$\dim_{\mathbb{C}} H^o(X, L_o^s \otimes L_1^t \otimes \ldots \otimes L_n^t) \leq m(sh_o(L) + th(L_1) + \ldots + th(L_n))^n$$

while the dimension of the space of polynomials $p(X_0, \ldots, X_n)$ of degree s in X_0, t in

$X_1, \ldots, X_n$ is $(s+1)(t+1)^n$. As soon as s is bigger than $m(h(L_1)+\ldots+h(L_n))$ one can always find t so that the dimension of the space of polynomials is greater than $\dim_{\mathbb{C}} H^o(X, L_o^{s} \otimes L_1^t \otimes \ldots \otimes L_n^t)$ whence a non-zero polynomial such that $p(f_0, \ldots, f_n)=0$. And the number s is independent of f_0, so we know that the field of meromorphic functions is an algebraic function field.

A variant form of this theorem is the following

III. I. THEOREM Let X be a compact, irreducible analytic space of dimension h, with holomorphic line bundle L. There is an integer B such that

$$\dim_C H^o(X, L^k) \leq B k^n$$

for all $k \geq 0$.

For the proof apply the lemma of the last theorem, noting that $h(L^k) \leq k\ h(L)$.

The theorem asserts that $\dim_{\mathbb{C}} H^o(X, L^k)$ grows like a polynomial in k, of degree not greater than the dimension of X. One can find compact spaces and line bundles where the degree of the polynomial is less than the dimension of X. For example on $\mathbb{P}^1 \times \mathbb{P}^1$ consider the line bundle $p_1^*(O_{\mathbb{P}^1}(1))$ pulled back from projection on the first factor, which has associated to it linear growth.

If $\varphi_1, \ldots, \varphi_j$ are global holomorphic sections of a line bundle L on X then $p_k(\varphi_1, \ldots, \varphi_j)$ is a global section of L^k on X, where p_k is any homogeneous polynomial in j variables of degree k. If this procedure yields the zero section of L^k just in case $p_k = 0$ then

$$\dim H^o(X, L^k) \geq \binom{j-1+k}{j-1} \sim \frac{k^{j-1}}{(j-1)!} \quad \text{for large } k$$

This must happen if, at some point, the meromorphic functions $\frac{\varphi_2}{\varphi_1}, \ldots, \frac{\varphi_j}{\varphi_1}$ are defined and give local coordinates. This happens, for example, with the line bundle $O_{\mathbb{P}}(1)$ on $\mathbb{P}^n$.

After having collected some more facts (mostly Chow's theorem, which we shall prove in the next section) we'll sketch another proof of Siegel's theorem. For now we'll discuss some examples relating to this theorem.

The main examples of compact analytic spaces with number of algebraically independent meromorphic functions equal to the dimension are the projective algebraic varieties. Here the transcendence degree of the field of rational functions is already equal to the dimension, so the field of meromorphic functions is at most a finite algebraic extension of that; we shall see later that the fields actually coincide.

A one dimensional compact analytic space always has algebraic dimension one and in fact is always a projective algebraic variety. A two dimensional compact analytic space might have algebraic dimension 0, 1, or 2. In case the space is a two dimensional complex manifold with algebraic dimension 2 then it is always a projective algebraic variety - although the corresponding statement is false in dimensions three or more, or in dimension two with singularieties. The simplest examples of compact complex manifolds with algebraic dimension less than geometric dimension are complex tori - spaces of the form $C^g/\mathcal{L}$, $\mathcal{L}$ a lattice - where the lattice does not satisfy Riemann's conditions - see Chern [8] or Kodaira and Morrow [19]. One gets a concrete example of a surface of algebraic dimension one by considering a certain complex structure on $S^3 \times S^1$, making it into what is called a Hopf surface. See Chern [8] for details of the construction. Topological considerations show that this is not an algebraic surface hence has algebraic dimension less than two. On the other hand there is a non-degenerate map from this manifold to $\mathbb{P}^1$ so it has some non-constant meromorphic functions.

Compact analytic spaces with algebraic dimension equal to geometric dimension are called Moisezon spaces. Moisezon has shown that they arise through modifications of algebraic varieties. They are important in algebraic geometry because most natural modifications of algebraic varieites can be performed in the category of Moisezon spaces and sometimes not in the category of algebraic varieties.

Chapter Three

§ 3 Chow's Theorem

Every algebraic subvariety of an algebraic variety defines an associated analytic subvariety. In general not every analytic subvariety of an algebraic variety arises in this way. The situation is different in the case of a projective algebraic variety. We shall prove Chow's theorem that every analytic subvariety of a projective algebraic variety is algebraic.

III. J. THEOREM Every analytic subvariety of $\mathbb{P}^N$ is algebraic.

It suffices to consider the case of an irreducible analytic $V \hookrightarrow \mathbb{P}^N$. Let $I \subset \mathbb{C}[X_0, \ldots, X_N]$ be the homogeneous ideal of polynomials vanishing on V. The algebraic variety $X = V^+(I) \supset V$ and it must be shown that there is equality. We know that I is prime so we know that X is algebraically irreducible. It will be enough to show that $\dim X = \dim V$ and X is analytically irreducible.

III. K. LEMMA Let X be an irreducible algebraic variety. The associated analytic space X_{an} is irreducible.

We consider first the case of X an irreducible algebraic hypersurface, $X = V(f) \hookrightarrow \mathbb{C}^n$. After a linear change of coordinates we may assume that $f(z_1, \ldots, z_n)$ is

$$z_n^d + p_{d-1}(z_1, \ldots, z_{n-1})z_n^{d-1} + \ldots + p_0(z_1, \ldots, z_{n-1}),$$

irreducible in the polynomial ring. The projection π of X onto the first n-1 coordinates is a finite analytic map with branch locus $B \subset \mathbb{C}^{n-1}$. For $z' \notin B$ there are locally defined holomorphic functions $x_1(z'), \ldots, x_d(z')$ which are permuted among themselves by analytic continuation throughout $\mathbb{C}^{n-1}-B$, the $x_i(z')$ being the roots of

$$z_n^d + p_{d-1}(z')z_n^{d-1} + \ldots + p_0(z').$$

If $\pi^{-1}(\mathbb{C}^{n-1}-B)$ is not connected then we can divide $\{x_1(z'),\ldots,x_d(z')\}$ in to two groups

$$\{x_1(z'),\ldots,x_a(z')\} \qquad \{x_{a+1}(z'),\ldots,x_d(z')\}$$

which are nonempty and never confused by analytic continuation. Setting

$\alpha_i = i+1^{st}$ elementary symmetric function of $x_1(z'),\ldots,x_a(z')$

$\beta_j = j+1^{st}$ elementary symmetric function of $x_{a+1}(z'),\ldots,x_d(z')$

the α_i's and β_j's are holomorphic and single-valued on $\mathbb{C}^{n-1}$-B and extend across B. There is a factorization

$$z_n^d + p_{d-1}(z')z_n^{d-1} + \ldots + p_0(z')$$

$$= (z_n^a + \alpha_{a-1}(z')z_n^{a-1} + \ldots + \alpha_0(z'))(z_n^{d-n} + \beta_{d-a+1}(z') + \ldots + \beta_0(z'))$$

and we would have a contradiction if all the α_i's, β_j's could be shown to be polynomials.

The following argument shows that the α_i's, β_j's have polynomial growth and are therefore polynomials. Consider $\mathbb{P}^{d+1}$ with homogeneous coordinates $(w, x_0,\ldots,x_d)$ and consider $\mathbb{C}^{d+1}$ imbedded as $D^+(w)$ with $z_i = \frac{x_i}{w}$ The strong topology closure of the hypersurface

$$V(z_d^d + z_d^{d-1}z_{d-1} + z_d^{d-2}z_{d-2} + \ldots + z_0) \longrightarrow \mathbb{C}^{d+1}$$

does not contain the point with homogeneous coordinates $(0,0,\ldots,0,1)$, which implies the existence of $\epsilon > 0$ such that

$$(1+|z_{d-1}| + |z_{d-2}| + \ldots + |z_0|) \geq \epsilon\, |z_d|$$

for all $(z_0,\ldots,z_d) \in V$. This gives a bound

$$(1+|p_{d-1}(z')|+\ldots+|p_0(z')|) \geq \epsilon |x_i(z')|$$

with ϵ independent of i, and this is what we wanted.

A consequence of this is that any Zariski open subset of an irreducible affine algebraic hypersurface is itself analytically irreducible. It is a fact from algebraic geometry that any irreducible algebraic variety contains a Zariski-dense Zariski open which is equivalent to an open on an irreducible hypersurface (because one can find a hypersurface with the same field of rational functions; see Lefschetz [20] or Šafarevič [35] for a proof). The general case of the lemma follows from this observation.

All we need show now is that dim X = dim V. The global sections of $\mathcal{O}_{\mathbb{P}^N}(1)_{hol}$ must give local coordinates at some point of V so we know that $\dim_{\mathbb{C}} H^0(V, \mathcal{O}_V(d))$ grows like a polynomial in d of degree m = dim V. In fact this argument shows that

$$\dim(H^0(\mathbb{P}^N, \mathcal{O}_{\mathbb{P}^N}(d)) / H^0(\mathbb{P}^N, I_V(d)))$$

grows like a polynomial of degree m in d. Since

$H^0(\mathbb{P}^N, I_V(d)) = H^0(\mathbb{P}^N, I_X(d))$ we fine that m = dim X.

Chow's theorem should be interpreted as a statement that analytic subsets of $\mathbb{C}^N$ which satisfy a certain growth condition must be algebraic. The growth condition in question is that the closure of each irreducible component in $\mathbb{P}^N$ again be analytic and irreducible.

There are some more comparison theorems which can be deduced from Chow's theorem. We'll give one:

III. L LEMMA Let $\varphi : X \longrightarrow X'$ be an algebraic map between algebraic varieties which is analytically an isomorphism. Then φ is algebraically an isomorphism.

We may assume that X (and therefore X') is irreducible. Since every point in X' has exactly one point of X mapped to it we know that φ is birational; this means that there is a proper algebraic subset D of X' on the complement of which φ inverts algebraically (see Lefschetz [20] or Šafarevič [35] for a discussion of birational maps). Then $\varphi^{-1}|_{X-D}$ is algebraic.

By locally exhibiting X' as a branched algebraic cover of some $\mathbb{C}^d$ and using symmetric functions we are reduced to proving that if $f: \mathbb{C}^d \longrightarrow \mathbb{C}$ is a holomorphic map the restriction of which to some non-empty Zariski open is algebraic then f is algebraic. But this is clear.

III. M COROLLARY Every holomorphic map between projective varieties is algebraic.

If $\varphi : X \longrightarrow Y$ is the map then the graph of φ is an algebraic subvariety of XxY (because the product of projective varieties is projective; see Šafarevič [35]). The projection graph $(\varphi) \longrightarrow X$ is algebraic with a holomorphic inverse. It is therefore an algebraic isomorphism and this proves the corollary.

Chapter Four GAGA

§ One

In this chapter we shall prove comparison theorems about sheaves and sheaf cohomology. The groundwork for this has already been laid in the computation of the holomorphic and algebraic cohomology of the sheaves $O_{\mathbb{P}^n}(d)$ on projective space.

If X is a projective algebraic variety we shall sometimes use the notations X_{alg} for X as a ringed space in the Zariski topology with sheaf O_{alg} and X_{hol} for X as an analytic space. There is a morphism of ringed spaces

$$\varphi : X_{hol} \longrightarrow X_{alg}$$

We know from sheaf theory that any coherent algebraic sheaf F will pull back to a coherent analytic sheaf $\varphi^*F = F_{hol}$ on X_{hol}. The stalk of this sheaf at a point $a \in X$ is $O_{a,hol} \otimes_{O_{a,alg}} F_a$. The coherence of F_{hol} is a consequence of the right-exactness of the tensor product.

For any i there are natural maps $H^i(X_{alg}, F) \longrightarrow H^i(X_{hol}, F_{hol})$. In general for a morphism of ringed spaces $f: X \longrightarrow Y$ and a sheaf F on Y there are induced functorial maps $H^i(Y, F) \longrightarrow H^i(X, f^*F)$. Maps $\check{H}^i(Y,F) \longrightarrow \check{H}^i(X, f^*F)$ are obviously defined and it suffices to consider these in the cases of our concern, since we can compute sheaf cohomology by the Čech method off an affine cover.

There are nice functorial properties for the association $F \longrightarrow F_{hol}$:

An O_{alg} linear map between coherent algebraic sheaves $F \longrightarrow G$ induces an O_{hol} linear map $F_{hol} \longrightarrow G_{hol}$.

Recall that there is a sheaf $\underline{Hom}_{O_{alg}}(F, G)$ the global sections of which are the

O_{alg}-module homomorphisms from F to G. There is a natural map of sheaves

$$(*) \qquad (\underline{\mathrm{Hom}}_{O_{alg}}(F, G))_{hol} \longrightarrow \underline{\mathrm{Hom}}_{O_{hol}}(F_{hol}, G_{hol})$$

which is an isomorphism. To see this recall that, for coherent sheaves F, G at a point $a \in X$

$$\underline{\mathrm{Hom}}_{O_{alg}}(F, G)_a \xrightarrow{\sim} \mathrm{Hom}_{O_{alg,a}}(F_a, G_a)$$

and similarly in the holomorphic case. Then at a point $a \in X$ the left side of (*) is

$$\mathrm{Hom}_{O_{alg,a}}(F_a, G_a) \otimes_{O_{alg,a}} O_{hol,a}$$

while the right side is

$$\mathrm{Hom}_{O_{hol,a}}(F_a \otimes_{O_{alg,a}} O_{hol,a}, G_a \otimes_{O_{alg,a}} O_{hol,a})$$

and the map is the natural one. We have reduced the question of the isomorphism of (*) - which it suffices to check at every point - to a question of pure algebra. The isomorphism can be deduced from the following algebraic facts

(1) For any exact sequence

$$F \longrightarrow G \longrightarrow H$$

of coherent algebraic sheaves the sequence

$$F_{hol} \longrightarrow G_{hol} \longrightarrow H_{hol}$$

of coherent analytic sheaves is exact.

(2) If F is a non-zero coherent algebraic sheaf then $F_{hol} \neq 0$.

These again reduce to questions at each stalk. Algebraically the point of all this is that for each $a \in X$ $O_{hol,a}$ is a faithfully flat extension of $O_{alg,a}$ - the stalk equivalents of (1), (2) serve as a definition of faithful flatness:

(1) For any exact sequence

$$A \longrightarrow B \longrightarrow C$$

of $O_{alg,a}$ modules the sequence

$$A \otimes_{O_{alg,a}} O_{hol,a} \longrightarrow B \otimes_{O_{alg,a}} O_{hol,a} \longrightarrow C \otimes_{O_{alg,a}} O_{hol,a}$$

is exact.

(2) If A is a non-zero $O_{alg,a}$ module then $A \otimes_{O_{alg,a}} O_{hol,a}$ is non-zero.

One proves that $O_{hol,a}$ if faithfully flat over $O_{alg,a}$ by noting that both are noetherian local rings with the same completions.

Since our point of view is analytic we shall not go into this, but only use these facts. For a discussion of the algebra involved see Altman-Kleiman [1] , or Matsumura [22] .

The comparison theorems which appear in this chapter are called the GAGA theorems, after the article of their first appearance, Serre [28] . The result can be stated in a single theorem:

THEOREM A *If* F *is a coherent analytic sheaf on a projective variety* X_{hol} *there is a unique coherent algebraic sheaf* F_{alg} *on* X_{alg} *such that* $F=(F_{alg})_{hol}$. *Furthermore the natural maps* $H^i(X_{alg}, F_{alg}) \longrightarrow H^i(X_{hol}, F)$ *are isomorphisms for all* i.

The key to GAGA is in the two theorems

THEOREM B *If* F *is a coherent analytic sheaf on* $\mathbb{P}^n_{hol}$ *there is a* d_0 *such that* $F(d) = F \otimes_{O_{\mathbb{P}^n_{hol}}} O_{\mathbb{P}^n_{hol}}(d)$ is generated by its global sections for $d \geq d_0$. — that is, *for each* $a \in \mathbb{P}^n$ *there are* $\{f_i\} \in H^0(\mathbb{P}^n_{hol}, F(d))$ *which generate* $F_a(d)$ *as an* $O_{\mathbb{P}^n_{hol,a}}$ *module.*

THEOREM C *If* F *is a coherent analytic sheaf on* $\mathbb{P}^n_{hol}$ *there is a* d_0 *such that* $H^i(\mathbb{P}^n_{hol}, F(d)) = 0$ *for* $i > 0,\ d \geq d_0$.

These two theorems are versions of Cartan's theorems A and B respectively, with growth conditions. Cartan's theorem A says that a coherent analytic sheaf on $\mathbb{C}^n$ is generated by its global sections and the first theorem says that in case this coherent sheaf extends across the hyperplane at infinity the generators can be chosen to have inessential singularities at infinity. The second theorem bears a similar relation to Cartan's theorem B. This is not to say that the proofs are directly derived from Cartan's theorems A and B. In fact the derivation of these two theorems will be fairly formal. The only analysis we'll use is a result about the finiteness of a cohomology group - although we need Cartan's theorem B to compute cohomology.

First note that it is sufficeint to prove GAGA for $X = \mathbb{P}^n$. For given an arbitrary projective algebraic $Y \longrightarrow \mathbb{P}^n$, we also have $Y_{hol} \longrightarrow \mathbb{P}^n_{hol}$, and a sheaf of $O_{Y_{alg}}$ (resp. $O_{Y_{hol}}$) - modules is coherent if and only if it is coherent as a sheaf of $O_{\mathbb{P}^n_{alg}}$ (resp. $O_{\mathbb{P}^n_{hol}}$) - modules. And the association $F \longrightarrow F_{hol}$ gives the same $O_{\mathbb{P}^n_{nol}}$ - module whether we consider F as an $O_{Y_{alg}}$ or $O_{\mathbb{P}^n_{alg}}$ module. Also the computation of cohomology groups can ignore this ambiguity.

Now we'll show how to get GAGA for $\mathbb{P}^n$ out of theorems B and C.

PROPOSITION D *Suppose that any coherent analytic sheaf* F *on* $\mathbb{P}^n$ *is a special* cokernel - that is, there is an exact sequence of coherent analytic sheaves

$$L_1 \longrightarrow L_0 \longrightarrow F$$

such that L_1, L_0 are of the form $O^{m_k}_{\mathbb{P}^n_{hol}}(\ell_k)$ for $k=0,1$. Then GAGA follows.

First of all, to get F_{alg} so that $F = (F_{alg})_{hol}$ the natural thing is to consider the morphism of algebraic sheaves

$$L_{1,alg} \longrightarrow L_{0,alg}$$

with $L_{k,alg} = O^{m_k}_{\mathbb{P}^n_{alg}}(\ell_k)$ and take F_{alg} = the cokernel.

The problem is to show that the map $L_1 \longrightarrow L_0$ is algebraic.

Now $\mathrm{Hom}_{O_{\mathbb{P}^n_{hol}}}(L_1, L_0) = H^0(\underline{\mathrm{Hom}}_{O_{\mathbb{P}^n_{hol}}}(L_1, L_0))$

$$\underline{\mathrm{Hom}}_{O_{\mathbb{P}^n_{hol}}}(L_1, L_0) = O^{m_1}_{\mathbb{P}^n_{hol}}(-\ell_1) \otimes L_0$$

$$= \otimes^{m_1}_{i=1} \otimes^{m_0}_{j=1} O_{\mathbb{P}^n_{hol}}(\ell_0 - \ell_1)$$

so that $\mathrm{Hom}_{O_{\mathbb{P}^n_{hol}}}(L_1, L_0)$ is $\otimes^{m_1}_{i=1} \otimes^{m_0}_{j=1} H^0(\mathbb{P}^n_{hol}, O_{\mathbb{P}^n_{hol}}(\ell_0 - \ell_1))$

and we know that this is the same as its algebraic counterpart. We can then let F_{alg} be the cokernel of

$$L_{1,alg} \longrightarrow L_{0,alg}$$

We know that

$$L_1 \longrightarrow L_0 \longrightarrow (F_{alg})_{hol} \longrightarrow 0$$

is exact and this ensures that $F = (F_{alg})_{hol}$ As for cohomology we'll show that the map

$$H^i(\mathbb{P}^n_{alg}, F_{alg}) \longrightarrow H^i(\mathbb{P}^n_{hol}, F)$$

is an isomorphism by descending induction on i. The following lemma starts the induction.

LEMMA E <u>If F is a coherent analytic or algebraic sheaf on</u> $\mathbb{P}^n$ <u>then</u> $H^i(\mathbb{P}^n, F) = 0$ <u>for</u> $i > n$.

This is because we can compute the group with alternating cochains off a cover with n+1 open sets.

We have the exact sequence

$$0 \longrightarrow A \longrightarrow L_{0,alg} \longrightarrow F_{alg}$$

where $A = \mathrm{Im}(L_{1,alg} \longrightarrow L_{0,alg})$, and the long exact cohomology sequence

$$0 \longrightarrow H^0(\mathbb{P}^n_{alg}, A) \longrightarrow H^0(\mathbb{P}^n_{alg}, L_{0,alg}) \longrightarrow H^0(\mathbb{P}^n_{alg}, F) \xrightarrow{\delta} H^1(\mathbb{P}^n_{alg}, A) \longrightarrow \cdots$$

and for each i maps

$$\begin{array}{ccccccccccc} \longrightarrow H^i(\mathbb{P}^n_{hol}, A_{hol}) & \longrightarrow & H^i(\mathbb{P}^n_{hol}, L_0) & \longrightarrow & H^i(\mathbb{P}^n_{hol}, F) & \xrightarrow{\delta} & H^{i+1}(\mathbb{P}^n_{hol}, A_{hol}) & \longrightarrow & H^{i+1}(\mathbb{P}^n_{hol}, L_0) \longrightarrow \\ (1)\uparrow & & (2)\uparrow & & (3)\uparrow & & (4)\uparrow & & (5)\uparrow \\ \longrightarrow H^i(\mathbb{P}^n_{alg}, A) & \longrightarrow & H^i(\mathbb{P}^n_{alg}, L_{0alg}) & \longrightarrow & H^i(\mathbb{P}^n_{alg}, F_{alg}) & \xrightarrow{\delta} & H^{i+1}(\mathbb{P}^n_{alg}, A) & \longrightarrow & H^{i+1}(\mathbb{P}^n_{alg}, L_{0alg}) \longrightarrow \end{array}$$

with the horizontal sequences exact and everything commuting.

(4), (5) are isomorphisms by inductive assumption and we know that (2) is an isomorphism. This implies that (3) is surjective. We get a similar result for the exact sequence

$$0 \longrightarrow B \longrightarrow L_{1\,alg} \longrightarrow A \longrightarrow 0$$

so we can conclude that (1) is surjective. The five lemma then shows that (3) is an isomorphism.

To show that a holomorphic sheaf is induced by a unique algebraic sheaf, suppose that F, G are coherent algebraic sheaves and $F_{hol} \xrightarrow{\sim} G_{hol}$. Then there is $\varphi \in \mathrm{Hom}_{Oalg}(F, G) = H^0(\underline{\mathrm{Hom}}_{Oalg}(F, G))$ inducing this isomorphism, and $\psi \in \mathrm{Hom}_{Oalg}(G, F)$ such that $\varphi \circ \psi = \mathrm{id}$ in $\mathrm{Hom}_{Ohol}(G_{hol}, G_{hol})$ $\psi \circ \varphi = \mathrm{id}$ in $\mathrm{Hom}_{Ohol}(F_{hol}, F_{hol})$. But then it must be that $\varphi \circ \psi = \mathrm{id}$, $\psi \circ \varphi = \mathrm{id}$ algebraically.

The proof of GAGA is thus reduced to showing that every coherent analytic sheaf is a special cokernel. Such information is provided by theorem B, according to which there is, for any F a coherent analytic sheaf and a d such that there is a surjective map

$$\oplus_1^m O_{\mathbb{P}^n}(d) \longrightarrow F \longrightarrow 0$$

which is a start. We do the same thing to the kernel of this map to get what we want.

Everything is now reduced to the proof of theorem B. We first show that theorem C is a consequence of theorem B.

The proof will be by descending induction on the order of the cohomology group, and we can again start the induction because $H^i(\mathbb{P}^n_{hol}, F) = 0$ for $i > n$. Represent F as a special cokernel

$$L_1 \longrightarrow L_0 \longrightarrow F \longrightarrow 0$$

Giving rise to the exact sequence

$$0 \longrightarrow A \longrightarrow L_0 \longrightarrow F \longrightarrow 0$$

There is d_0 such that for $d \geq d_0$, $H^i(\mathbb{P}^n, L_0(d)) = 0$, $H^{i+1}(\mathbb{P}^n, A(d)) = 0$, and the exact sequence

$$H^i(\mathbb{P}^n, L(d)) \longrightarrow H^i(\mathbb{P}^n, F(d)) \longrightarrow H^{i+1}(\mathbb{P}^n, A(d))$$

shows that

$$H^i(\mathbb{P}^n, F(d)) = 0 \text{ for } d \geq d_0.$$

We'll need this in our proof of theorem B, which is by induction on the dimension n of the projective space. We start with $\mathbb{P}^0$ - where there is nothing to prove.

LEMMA F <u>For a coherent analytic sheaf</u> F <u>on</u> $\mathbb{P}^n$ <u>there is</u>, <u>for each</u> $a \in \mathbb{P}^n$, <u>a</u> d_0 <u>such that the stalk</u> $F_a(d)$ <u>is generated by the global sections</u> $H^0(\mathbb{P}^n_{hol}, F(d))$ <u>for all</u> $d \geq d_0$.

Pick a hyperplane $\mathbb{P}^{n-1}$ of $\mathbb{P}^n$ passing through a. The ideal sheaf of this is isomorphic to $O_{\mathbb{P}^n_{hol}}(-1)$, and we fix a map $O_{\mathbb{P}^n_{hol}}(-1) \longrightarrow O_{\mathbb{P}^n_{hol}}$. For all d there is the exact sequence

$$0 \longrightarrow F(d-1) \longrightarrow F(d) \longrightarrow F(d)\big|_{\mathbb{P}^{n-1}} \longrightarrow 0$$

which gives the cohomology exact sequence

$$0 \longrightarrow H^0(\mathbb{P}^n, F(d-1)) \longrightarrow H^0(\mathbb{P}^n, F(d)) \longrightarrow H^0(\mathbb{P}^{n-1}, F(d)) \longrightarrow H^1(\mathbb{P}^n, F(d-1)) \longrightarrow$$

$$\longrightarrow H^1(\mathbb{P}^n, F(d)) \longrightarrow H^1(\mathbb{P}^{n-1}, F(d)) \longrightarrow \ldots$$

By inductive hypothesis there is d_1 such that $H^1(\mathbb{P}^{n-1}, F(d)) = 0$ for $d \geq d_1$, so that the map $H^1(\mathbb{P}^n, F(d-1)) \longrightarrow H^1(\mathbb{P}^n, F(d))$ is surjective for $d \geq d_1$.

If we let d grow we get a long sequence.

$$H^1(\mathbb{P}^n, F(d_1-1)) \longrightarrow H^1(\mathbb{P}^n, F(d_1)) \longrightarrow H^1(\mathbb{P}^n, F(d_1+1)) \longrightarrow \ldots \longrightarrow H^1(\mathbb{P}^n, F(d)) \longrightarrow \ldots$$

with the maps surjective at each stage.

THEOREM F <u>The cohomology groups of a coherent analytic sheaf on a compact analytic space are finite dimensional over $\mathbb{C}$.</u>

This theorem is proved in Gunning-Rossi [13]. It implies that there is some $d_2 \geq d_1$ such that the maps $H^1(\mathbb{P}^n, F(d)) \longrightarrow H^1(\mathbb{P}^n, F(d+1))$ are isomorphisms for $d \geq d_2$ so

$$0 \longrightarrow H^0(\mathbb{P}^n, F(d-1)) \longrightarrow H^0(\mathbb{P}^n, F(d)) \longrightarrow H^0(\mathbb{P}^{n-1}, F(d)|_{\mathbb{P}^{n-1}}) \longrightarrow 0$$

is exact for $d \geq d_2$. Pick $d_0 \geq d_2$ so that the global sections of $H^0(\mathbb{P}^{n-1}, F(d)|_{\mathbb{P}^{n-1}})$ generate for $d \geq d_0$. Then the elements of $H^0(\mathbb{P}^n, F(d))$ generate $F(d)_a$ over $O_{\mathbb{P}^n_a}$ - modulo the ideal defining $\mathbb{P}^{n-1}$ at a. It follows from Nakayama's lemma that the global sections $H^0(\mathbb{P}^n, F(d))$ generate $F(d)_a$. for $d \geq d_0$.

Theorem B follows by a compactness argument from this lemma.

COROLLARY G (Chow's theorem) <u>Every analytic subvariety of a projective variety is algebraic.</u>

If X is an analytic subvariety of the projective algebraic variety V then it is the support of a coherent analytic sheaf and thus the support of a coherent algebraic sheaf.

COROLLARY H <u>Every holomorphic vector bundle on a projective variety is induced by a unique algebraic vector bundle.</u>

It must be shown that if F is a coherent algebraic sheaf and F_{hol} is locally free then F is locally free. This reduces to a local statement and follows from flatness.

For our later purposes this last corollary is the most important of the GAGA results. It says that a holomorphic vector bundle on an affine variety which satisfies a growth condition, to the effect of extending across the section at infinity of the variety, must be algebraic. We mention another interesting corollary, the proof of which is contained in previous remarks.

COROLLARY I. Every holomorphic line bundle on a projective variety is the line bundle of a divisor.

CHAPTER FIVE

§ 1 Line Bundles, Divisors, and Maps to $\mathbb{P}^n$

Every complex manifold has a natural orientation, so that on a compact complex manifold M of dimension n there is defined a prefered generator of $H_{2n}(M, \mathbb{Z})$.

If D is a complex submanifold of codimension one of the compact complex manifold M then the image of the prefered generator of $H_{2n-2}(D, \mathbb{Z})$ defines a class $[D]$ in $H_{2n-2}(M, \mathbb{Z})$ and by Poincaré duality a class $[D]$ in $H^2(M, \mathbb{Z})$. This is the cohomology class of the divisor.

If M is a possibly non-compact complex manifold and D is a divisor, possibly with singularities, we can still define $[D] \in H^2(M, \mathbb{Z})$, as we shall see later in this section. In the last section of this chapter we shall show how to define the cohomology class of any analytic subvariety of a complex manifold. Our study of the resulting analytic cohomology classes, or analytic cocycles, will lean heavily on the theory of vector bundles. In this section we shall discuss these ideas in the special case of divisors and line bundles.

Let M be a complex manifold of dimension n, D an effective divisor on M. There is a holomorphic line bundle $L \longrightarrow M$, with a holomorphic section corresponding to D. There is an exact sequence of sheaves on M

$$o \longrightarrow \mathbb{Z} \longrightarrow O \xrightarrow{\exp 2\pi\sqrt{-1}} O^x \longrightarrow 1$$

where $\mathbb{Z}$ denotes the sheaf associated to the constant presheaf with stalk $\mathbb{Z}$. There is induced a coboundary map

$$\delta\colon H^1(M, O^x) \longrightarrow H^2(M, \mathbb{Z}).$$

Since M is a manifold the second cohomology of M with coefficients in $\mathbb{Z}$ is the same as the second singular cohomology group of M with $\mathbb{Z}$ coefficients. The kernel of the map δ is the image of $H^1(M, O) \longrightarrow H^1(M, O^x)$. For a holomorphic line bundle L, $\delta(L)$ is called the <u>first Chern class</u> of L and denoted $c_1(L)$.

One can do the same thing with arbitrary differentiable complex line bundles on M:

$$\begin{array}{ccccccccc} o & \longrightarrow & \mathbb{Z} & \longrightarrow & O_{diff} & \xrightarrow{\exp 2\pi\sqrt{-1}} & O^x_{diff} & \longrightarrow & 1 \\ & & \downarrow & & \downarrow & & \downarrow & & \\ o & \longrightarrow & \mathbb{Z} & \longrightarrow & O & \xrightarrow{\exp 2\pi\sqrt{-1}} & O^x & \longrightarrow & 1 \end{array}$$

$\delta: H^1(M, O^x_{diff}) \longrightarrow H^2(M, \mathbb{Z})$ assigns to each differentiable complex line bundle its first Chern class. The diagram shows that the first Chern class of a holomorphic bundle depends only on its differentiable structure. Furthermore, $H^1(M, O_{diff}) = H^2(M, O_{diff}) = 0$ because these sheaves are flabby. Then

$$H^1(M, O^x_{diff}) \xrightarrow{\sim} H^2(M, \mathbb{Z})$$

so that $H^2(M, \mathbb{Z})$ on a complex manifold may be identified with the group of complex differentiable line bundles.

THEOREM A <u>Let</u> D <u>be a smooth divisor on the smooth, projective variety</u> M, <u>with holomorphic line bundle</u> $L \longrightarrow M$. <u>The cohomology class of</u> D <u>agrees with the first Chern class of</u> L.

The proof of this theorem will require ideas to be developed in the rest of this section. The theorem is actually true in the generality of any divisor on a complex manifold, although we shall not prove that here.

The proof will depend on the introduction of a classifying space for complex line bundles, which we shall discuss now.

The most important line bundle in algebraic geometry is the line bundle $O_{\mathbb{P}^n}(1)$ on projective space. The divisors associated to $O_{\mathbb{P}^n}(1)$ are the linear hyperplanes $\mathbb{P}^{n-1} \longrightarrow \mathbb{P}^n$. There is a canonical isomorphism $H^2(\mathbb{P}^1, \mathbb{Z}) \xrightarrow{\sim} \mathbb{Z}$, and any linear inclusion $\mathbb{P}^1 \longrightarrow \mathbb{P}^n$ induces an isomorphism $H^2(\mathbb{P}^n, \mathbb{Z}) \xrightarrow{\sim} H^2(\mathbb{P}^1, \mathbb{Z})$ giving a canonical generator for $H^2(\mathbb{P}^n, \mathbb{Z})$. (For facts on the topology of $\mathbb{P}^n$, see Spanier [29]).

PROPOSITION B *On* $\mathbb{P}^n$,

$$c_1(O_{\mathbb{P}^n}(1)) = \text{positive generator of } H^2(\mathbb{P}^n, \mathbb{Z})$$

$$= \text{cohomology class of a hyperplane.}$$

The proof is by induction on n. Since $O_{\mathbb{P}^n}(1)\Big|_{\mathbb{P}^{n-1}} = O_{\mathbb{P}^n}(1)$, and since taking Chern classes commutes with restriction, it suffices to prove this for $\mathbb{P}^1$. But this is obvious.

Now consider continuous complex line bundles on finite polyhedra. The complex projective spaces are classifying spaces for the funtor which assoicates to each finite polyhedron its group of complex line bundles. This means that, given a complex line bundle L on a CW complex X of dimension $\leq 2n$, there is a map $f: X \longrightarrow \mathbb{P}^n$ such that $L \xrightarrow{\sim} f^*(O_{\mathbb{P}^n}(1))$. The map f is unique up to homotopy. Forming the limit as a topological space $\mathbb{P}^\infty = \lim_n \mathbb{P}^n$, with the line bundle $O_{\mathbb{P}^\infty}(1) = \lim_n O_{\mathbb{P}^n}(1)$ we can state the following

THEOREM C *For any polyhedron* X *there is a* 1-1 *correspondence between isomorphism classes of continuous complex line bundles on* X *and homotopy classes of maps from* X *to* $\mathbb{P}^\infty$.

A proof of this theorem appears in Spanier [29]. We shall prove a stronger theorem with this as a corollary later in this chapter.

Consider now this theorem in the special case of holomorphic line bundles on analytic spaces. Suppose the holomorphic line bundle L on the analytic space X to be generated by a finite number of global sections, $\varphi_o, \ldots, \varphi_n \in H^o(X, L)$. By picking a trivialization of L around any point a, say on a neighborhood U of a, one gets functions $\varphi_{o,U}, \ldots, \varphi_{n,U}$ which do not vanish simultaneously, and from these one gets a holomorphic map $U \longrightarrow \mathbb{C}^{n+1} - \{0\}$ which in turn defines a holomorphic map $U \longrightarrow \mathbb{P}^n$. The last map does not depend on the trivialization and in this way a holomorphic map $X \xrightarrow{\varphi} \mathbb{P}^n$ has been defined.

This map induces an isomorphism $L \xrightarrow{\sim} \varphi^*(O_{\mathbb{P}^n}(1))$ with $\varphi_o, \ldots, \varphi_n$ as the pull-backs of the global sections $x_o, \ldots, x_n$ of $O_{\mathbb{P}^n}(1)$ on $\mathbb{P}^n$. Conversely any holomorphic map $\varphi\colon X \longrightarrow \mathbb{P}^n$ defines a holomorphic line bundle $\varphi^*(O_{\mathbb{P}^n}(1))$ generated by the global sections $\varphi^*(x_o), \ldots, \varphi^*(x_n)$. This proves the

THEOREM D <u>There is a one-to-one correspondence between holomorphic maps from an analytic space X to $\mathbb{P}^n$ and holomorphic line bundles on X together with n+1 global sections generating each fibre.</u>

A corollary of this theorem is that every holomorphic automorphism of $\mathbb{P}^n$ is induced by a linear automorphism of $\mathbb{C}^{n+1}$, so that $\mathrm{Aut}(\mathbb{P}^n) \xrightarrow{\sim} PGL(n+1, \mathbb{C}) = GL(n+1, \mathbb{C})/\mathbb{C}^x$.

Also the proof of this theorem suggests how, on an arbitrary polyhedron with a complex line bundle, to construct a map to $\mathbb{P}^\infty$. It's a problem of finding global sections which generate.

Theorem D has an analogue for algebraic line bundles on algebraic varieties, proved in exactly the same way.

THEOREM E There is a one-to-one correspondence between algebraic maps from an algebraic variety X to $\mathbb{P}^n$ and algebraic line bundles on X together with $n+1$ global sections generating each fiber.

By GAGA (or Chow's theorem) theorems D and E are equivalent in the projective case.

It is important to know when the map to $\mathbb{P}^n$ which a line bundle induces on an analytic space is an imbedding. Suppose then that a holomorphic line bundle L on the space X has global sections $\varphi_o, \ldots, \varphi_n$ which generate, inducing the map $\varphi : X \to \mathbb{P}^n$. We'll suppose for simplicity that the φ_i are a basis of $H^o(X, L)$. φ will be an isomorphism in a neighborhood of a point $a \in X$ just in case it induces an isomorphism of the tangent space to X at a onto its image under the map $d\varphi_a$. This translates in sheaf-theoretic terms into the requirement that the map

$$H^o(X, L \otimes I_a) \to H^o(X, L \otimes I_a/I_a^2)$$

induced by

$$o \to L \otimes I_a^2 \to L \otimes I_a \to L \otimes I_a/I_a^2 \to 0$$

where I_a is the sheaf of ideals of functions vanishing at a_j is surjective.

If φ satisfies this condition and is a local isomorphism it will define an imbedding just in case it separates points. Two points $a, b \in X$ will be separated if one can find an element of $H^o(X, L)$ which vanishes at a but not at b. In sheaf-theoretic terms it is required that

$$o \to L \otimes I_{a \cup b} \to L \otimes I_a \to L \otimes O/I_b \to 0$$

induce a non-zero map

$$H^o(X, L\otimes I_a) \longrightarrow H^o(X, L\otimes \mathcal{O}/I_b)$$

If follows that the line bundle L will define an imbedding into $\mathbb{P}^n$ is the two conditions

$$H^1(X, L\otimes I_{a\cup b}) = 0$$

$$H^1(X, L\otimes I_a^2) = 0$$

are satisfied for all a, b in X.

A holomorphic line bundle which defines an imbedding is called <u>very ample</u>. A holomorphic line bundle is said to be <u>ample</u> if some positive power of it is very ample.

The prototypical ample line bundle is then $O_{\mathbb{P}^n}(1)\big|_X$ for some $X \longrightarrow \mathbb{P}^n$. The sections of $O_{\mathbb{P}^n}(1)$ over $\mathbb{P}^n$ will define divisors on X which are the intersections of X with hyperplanes of $\mathbb{P}^n$. These are called <u>hyperplane sections.</u> The hyperplanes of $\mathbb{P}^n$ are put in one-to-one correspondence with the points of $\mathbb{P}^n$ by the choice of an inner product in $\mathbb{C}^{n+1}$, so it is possible to speak of an algebraic family of hyperplanes.

In the traditional language of algebraic geometry, a property which all points of an algebraic variety possess, except perhaps the points of a proper algebraic subvariety, is said to be enjoyed by the <u>generic</u> point of the variety. In this language we state the

THEOREM F (Bertini's theorem) <u>The generic hyperplane section of a non-singular, projective variety is a non-singular subvariety.</u>

Let $X \longrightarrow \mathbb{P}^n$ be the smooth variety of dimension d. We'll denote the space of hyperplanes in $\mathbb{P}^n$ by $\mathbb{P}^{n*}$. It will be enough to show that the generic element of $\mathbb{P}^{n*}$ is not tangent to X.

Let V be the subvariety of $X \times \mathbb{P}^{n*}$ consisting of pairs (x, H) such that H is tangent to X at x - that is, H contains all lines tangent to X at x. Since the tangent space to each $x \epsilon X$ is d dimensional the fibre of the map $V \longrightarrow X$ consists of all hyperplanes containing a d dimensional subspace of $\mathbb{P}^n$. By duality this is the same as the points contained in a n-d-1 dimensional subspace of $\mathbb{P}^n$ and so has dimension n-d-1. Then in the map $V \longrightarrow D$ the fiber has dimension n-d-1 and the base has dimension d, so $\dim V = n-1$.

We use a fact from algebraic geometry: the image under an algebraic map of a projective, irreducible variety of dimension k is a subvariety of dimension less than or equal to k. See Šafarevič [35].

The projection of V to $\mathbb{P}^{n^*}$ has a proper algebraic image, and this is what we wanted.

Another theorem on hyperplane sections is the

THEOREM G *Suppose that* S *is a non-singular subvariety of codimension one on the non-singular variety* X *and that the line bundle associated to* S *is ample. Suppose that* $\dim X = n$. *The the map*

$$H_q(S, \mathbb{Z}) \longrightarrow H_q(X, \mathbb{Z})$$

is an isomorphism for $q < n-1$ *and is onto for* $q = n-1$. *In particular* S *is connected (and theorefore irreducible).*

This is a theorem of Lefschetz. We shall not prove it here. A proof appears in the article of Bott [5].

We shall use the Bertini and Lefschetz theorems in our proof of theorem A.

PROPOSITION H *Let* D_1, D_2 *and* B *be distinct smooth, irreducible divisors on the smooth, projective algebraic variety* M. If $D_1 + D_2$ *is linearly equivalent to* B *then there is an equality of cohomology classes* $[B] = [D_1] + [D_2]$.

By intersecting everything with suitable hyperplane sections, using the Lefschetz theorem to get injections on cohomology and iterating this procedure, we see that it is enough to consider the case of curves on an algebraic surface.

Consider first the case where D_1, D_2, and B are all disjoint. Then there will be sections G_1, G_2 of the line bundle L such that the zeroes of G_1 are D_1 and D_2 while the zeroes of G_2 are B. The meromorphic function $\frac{G_1}{G_2}$ gives an everywhere defined holomorphic map $f: M \longrightarrow \mathbb{P}^1$ with $f^{-1}(o) = D_1 + D_2$ and $f^{-1}(\infty) = B$. By taking a path from o to ∞ on $\mathbb{P}^1$ which avoids the finitely many critical values of f we get a real submanifold of M with boundary the disjoint union of D_1, D_2 and B. It follows now from the topology of this situation that $[B] = [D_1] + [D_2]$ (see Stong [40]).

In the general case there will be a finite number of points where B will intersect either D_1 or D_2. It is a fact from algebraic geometry that we can find a new surface N and an algebraic map $g: N \longrightarrow M$ such that

(1) the map $H_2(N, \mathbb{Z}) \longrightarrow H_2(M, \mathbb{Z})$ will be surjective

(2) D_1, D_2, and B will be imbedded as smooth divisors on N, so that B will be linearly equivalent to $D_1 + D_2$. B will be disjoint from D_1 and D_2.

(3) the maps $D_i \longrightarrow M$, $B \longrightarrow M$ factor through $D_i \longrightarrow N_1$ $B \longrightarrow N$

N will be constructed by <u>blowing up points</u> in M. See Šafarevič [29].

The proposition now almost follows from the special case. It is enough to show that $[B] = [D_1] + [D_2]$ on N. Here there will again be a holomorphic map $f: M \longrightarrow \mathbb{P}^1$ with $D_1 + D_2 = f^{-1}(o)$, $B = f^{-1}(\infty)$. Only if D_1 and D_2 intersect o may be a critical value for f. But it will always be possible, by taking a path from o to ∞ avoiding all possible critical values except o, then moving the resulting set a little bit, to get a real manifold with boundary mapping into N, with boundary in pieces D_1, D_2 and B. Again this implies that $[B] = [D_1]+[D_2]$.

From the characterization of ample line bundles in terms of sheaf cohomology we get the

PROPOSITION I *Suppose that* $X \longrightarrow \mathbb{P}^n$ *is a non-singular projective variety. For any holomorphic line bundle* L *on* X *there is a* d_o *such that* $L(d) = L \otimes_{O_X} O_{\mathbb{P}^n}(d)$ *is very ample for* $d \geq d_o$.

Now we complete the proof of theorem A: If D is a smooth divisor with line bundle L on the projective variety M we can assume that M is projectively imbedded so that $L \otimes_{O_M} O_{\mathbb{P}^n}(1)$ is very ample. If H is a hyperplane section we know that the cohomology class of H is the Chern class of $O_{\mathbb{P}^n}(1)$ restricted to M; also if B is a smooth, irreducible divisor of the line bundle L(1) then the cohomology class of B is the first Chern class of L(1), which is $c_1(L) + c_1(O_{\mathbb{P}^n}(1)))_M$. Theorem A follows by subtraction.

A consequence of this discussion is the

THEOREM J *The subgroup of* $H^2(M, \mathbb{Z})$ *given by Chern classes of holomorphic line bundles is generated by the cohomology classes of smooth divisors.*

Summarizing the results stated so far in this section, algebraic subvarieties of codimension one have been represented as the zeroes of sections of line bundles, sometimes as the hyperplane sections of projective embeddings. The cohomology class of such a subvariety is represented by the Chern class of the line bundle associated to the divisor.

A word should be inserted here about the representation of the Chern class by differential forms, after mapping $H^2(X, \mathbb{Z}) \longrightarrow H^2(X, \mathbb{C})$, with X a complex manifold. Since this will be generalized later no proofs will be given.

First the notion of a hermitian metric on a vector bundle must be introduced. Consider a continuous vector bundle F over a topological space X, with fiber $\mathbb{C}^n$, trivialized by the covering $\{U_i\}$ with transition functions $f_{ij}: U_i \cap U_j \longrightarrow GL(n, \mathbb{C})$. One knows what is meant by a Hermitian metric on the vector space $\mathbb{C}^n$: It is given by a positive definite Hermitian matrix $h \in GL(n, \mathbb{C})$. Accordingly a hermitian metric on the trivial bundle $F|U_i$ is given by a continuous map $h_i: U_i \longrightarrow GL(n, \mathbb{C})$ taking positive hermitian matrices as values. To get a hermitian metric on F such things must patch together. What is needed is a set of maps $h_i: U_i \longrightarrow Herm^+(n, \mathbb{C})$ such that ${}^t\bar{f}_{ij} h_i f_{ij} = h_j$ on $U_i \cap U_j$. Introducing a Hermitian metric on a complex vector bundle amounts to reducing the structure group from $GL(n, \mathbb{C})$ to the unitary group $U(n, \mathbb{C})$.

THEOREM K <u>Every vector bundle on a paracompact topological space admits a Hermitian metric.</u>

This is an argument using partitions of unity; it will not be given here. A proof appears in Hormander [17].

This theorem applies in particular to a 2nd countable manifold. Also in the case of a differentiable vector bundle on a 2nd countable manifold the same proof guarantees the existence of a differentiable Hermitian metric.

If L is a holomorphic line bundle on a complex manifold X, trivialized by $\{U_i\}$ and with transition functions $\alpha_{ij}: U_i \cap U_j \longrightarrow \mathbb{C}^x$, then picking a Hermitian metric on L amounts to picking differentiable $a_i: U_i \longrightarrow \mathbb{R}^+$ such that

$$|\alpha_{ij}|^2 a_i = a_j \quad \text{on} \quad U_i \cap U_j$$

Because the transition functions α_{ij} are holomorphic there is an equality of 2-forms

$$\bar{\partial}\partial \log a_i = \bar{\partial}\partial \log a_j \qquad \text{on} \quad U_i \cap U_j$$

THEOREM L *The 2-form defined on U_i by*

$$\frac{1}{2\pi\sqrt{-1}} \bar{\partial}\partial \log a_i$$

is closed and the cohomology class if represents in $H^2(X, \mathbb{C})$ is $c_1(L)$.

A more general form of this theorem will be proven in the next section.

CHAPTER FIVE

§2 Grassmannians and Vector Bundles

The first step in the generalization of the results of the last section is in the definition of the Grassmann varieties, which are to higher dimensional vector bundles as the projective spaces are to line bundles. For a given $k \leq n$, consider $\mathbb{C}^{kn}$ as the space of all $k \times n$ matrices. The subset S of all matrices of rank less than k is an algebraic subset, so $\mathbb{C}^{kn}-S$ is a Zariski open set $GL(k,\mathbb{C})$ acts by left multiplication on $\mathbb{C}^{kn}-S$, and the quotient topological space by this action is called Grass(k, n), the Grassmannian of all k-planes in n-space. In case $k=1$ the construction of projective space has been repeated. Exactly as in the construction of projective space a sheaf of rings is defined on Grass(k, n) which makes it into a complex manifold. Also one could repeat the construction from an algebraic point of view in the Zariski topology, which shows that Grass(k, n) is the complex manifold associated to an algebraic variety.

There is the holomorphic map

$$\phi : \mathbb{C}^{kn} - S \longrightarrow \text{Grass}(k,n)$$

Consider the Zariski open U on $\mathbb{C}^{kn}-S$ where the first k columns of the matrix are linearly independent. U is invariant under the action of $GL(k,\mathbb{C})$ so $\phi(U)$ is an open set in Grass (k, n). U is naturally isomorphic to

$$GL(k,\mathbb{C}) \times \mathbb{C}^{k(n-k)}$$

with the $GL(k,\mathbb{C})$ action as left multiplication, and the quotient of U may be identified with $\mathbb{C}^{k(n-k)} \xrightarrow{\simeq} \phi(U)$. A similar analysis in the case where the columns $i_1, \ldots, i_k$ (with $i_1 < \ldots < i_k$) are linearly independent shows that Grass (k, n) is covered by open sets $W_{i_1 \ldots i_k} = \phi(U_{i_1 \ldots i_k})$, each isomorphic, both holomorphically and as algebraic varieties, to $\mathbb{C}^{k(n-k)}$.

One says that

$$\phi : \mathbb{C}^{kn} - S \longrightarrow \text{Grass}(k, n)$$

is a principal fiber bundle, both holomorphically and algebraically, with fiber $GL(k, \mathbb{C})$.

$\mathbb{C}^{kn} - S$ may be thought of either as the set of k-tuples of linearly independent vectors in $\mathbb{C}^n$ or as the set of all surjective maps from $\mathbb{C}^n$ to $\mathbb{C}^k$. Thus Grass (k, n) may be thought of either as the set of k-dimensional linear subspaces of $\mathbb{C}^n$ or as the set of surjective maps from $\mathbb{C}^n$ to $\mathbb{C}^k$, modulo isomorphisms of $\mathbb{C}^k$. Thinking of the Grassmannian in terms of subspaces, denote $\mathbb{C}^k - S$ by St(k, n), the Stiefel manifold of k-frames in $\mathbb{C}^n$, which is to say of k-tuples of linearly independent vectors in C^n. Then the map $\phi : St(k, n) \longrightarrow \text{Grass}(k, n)$ assigns to each k-frame the subspace it spans. Note that Grass (1, n) is $\mathbb{P}^{n-1}$ and that Grass (k, n) may be thought of as the space of $\mathbb{P}^{k-1}$'s in $\mathbb{P}^{n-1}$.

Thinking of Grass(k, n) as k-spaces in $\mathbb{C}^n$, it can be represented in another way. $GL(n, \mathbb{C})$ acts transitively on the k-spaces. We will denote by $GL(k, n-k, \mathbb{C})$ the subgroup which leaves fixed the k-space $z_{k+1} = \ldots = z_n = 0$. The Grassmannian manifold may be identified, as a complex manifold on algebraic variety, with $GL(n, \mathbb{C}) / GL(k, n-k, \mathbb{C})$.

This shows that Grass (k, n) has a transitive group of algebraic automorphisms. In terms of our previous discussion, $GL(n, \mathbb{C})$ acts by right multiplication on St(k, n) and this action descends to Grass (k, n).

Fixing the usual Hermitian inner product in $\mathbb{C}^n$, the unitary group $U(n, \mathbb{C})$ also acts transitively on the k-spaces. Denoting the subgroup leaving the space defined by $z_{k+1} = \ldots = z_n = 0$ fixed by $U(k, n-k, \mathbb{C})$, the Grassmannian may be identified, as differ -entiable manifold, with $U(n, \mathbb{C}) / U(k, n-k)$. This shows that the Grassmannian is compact.

The Grassmannians are in fact projective, and each has a special projective imbedding, called the Plücker imbedding.

To get this imbedding, first map $St(k, n) \longrightarrow \mathbb{P}^{\binom{n}{k}-1}$ as follows : Use homogeneous coordinates $x_{i_1 \ldots i_k}$, $1 \leq i_1 < i_k \leq n$, in $\mathbb{P}^{\binom{n}{k}-1}$, and map a point of $\mathbb{C}^{kn} - S$ to the determinant of the $i_1, \ldots, i_k$ columns. The action of $GL(k, \mathbb{C})$ on $St(k, n)$ will change the homogeneous coordinates by a constant multiple, so there is an algebraic and holomorphic map

$$\text{Grass}\,(k, n) \xrightarrow{p} \mathbb{P}^{\binom{n}{k}-1} .$$

We will show that p is an imbedding. First consider p restricted to $\phi(U_{1 \ldots k}) = \phi(U)$. This can be identified with $\mathbb{C}^{k(n-k)}$, and p maps it into $\mathbb{C}^{\binom{n}{k}-1} = D^+(x_{1 \ldots k})$ by thinking of $\mathbb{C}^{k(n-k)}$ as the set of all kxn matrices with first k columns the identity matrix

$$\widetilde{A} = (I_k A)$$

and $p_{i_1 \ldots i_k}(A) =$ determinant of columns $i_1, \ldots, i_k$ of $\widetilde{A}$. Since $p_{1 \ldots i \ldots k\, k+j}(A) = i+j^{\text{th}}$ coordinate of A, this shows that p is an imbedding restricted to $\phi(U)$. One can do the same thing on the opens $\phi(U_{i_1 \ldots i_k})$, from which it follows that p is an imbedding everywhere.

The polynomials defining the Grassmannian under the Plücker imbedding may be written down explicitly. In $\mathbb{P}^{\binom{n}{k}-1}$, again we use homogeneous coordinates $x_{i_1 \ldots i_k}$ for every k-tuple $1 \leq i_1 < i_2 < \ldots < i_k \leq n$. For any k-tuple of numbers between 1 and n, $i_1, \ldots, i_k$, let $x_{i_1 \ldots i_k}$ be zero if two of the indices are the same; otherwise let it equal $\epsilon(\sigma) x_{i_{\sigma(1)} \ldots i_{\sigma(k)}}$, where $\sigma: \{1, \ldots, k\} \longrightarrow \{1, \ldots, k\}$ is the permutation such that $i_{\sigma(1)} < i_{\sigma(2)} < \ldots < i_{\sigma(k)}$.

For any pair of k-tuples $1 \leq i_1 < i_2 < \ldots < i_k \leq n$, $1 \leq j_1 < \ldots < j_k \leq n$ and any $\alpha \in \{1, \ldots, k\}$ there is a homogeneous polynomial

$$R_{(i_1 \ldots i_k)(j_1 \ldots j_k)(\alpha)} =$$

$$x_{i_1 \ldots i_k} x_{j_1 \ldots j_k} + \Sigma_{q=1}^{k} x_{i_1 \ldots \hat{i}_\alpha j_q i_{\alpha+1} \ldots i_k} x_{j_1 \ldots j_q i \ldots j_k} (-1)^q$$

The Grassmannian is the variety defined by the $R_{i,j,\alpha}$.

To prove that the Grassmannian satisfies these relations it suffices to consider the case $i_1, \ldots, i_k = 1, 2, \ldots, k-1, k$ and to show that on $W = \phi(U_{1 \ldots k})$ the map

$$p : \mathbb{C}^{k(n-k)} \longrightarrow \mathbb{C}^{\binom{n}{k}-1} = D^+(x_{1 \ldots k})$$

has image contained in the zero locus of

$$\tilde{R}_{1, \ldots, k, j, \alpha} = t_{j_1 \ldots j_k} + \Sigma_{q=1}^{k} t_{1 \ldots j_q \ldots k} t_{j_1 \ldots \hat{j}_q \ldots j_k} (-1)^q$$

α^{th} place

where affine coordinates $t_{j_1 \ldots j_k} = \dfrac{x_{j_1 \ldots j_k}}{x_{1 \ldots k}}$ are used for $\mathbb{C}^{\binom{n}{k}-1}$.

Again consider an element A of $\mathbb{C}^{k(n-k)}$ as a $k\mathrm{x}(n-k)$ matrix, letting $\widetilde{A} = (I_k A)$ and $p_{j_1 \dots j_k}(A) = \det$ of columns $j_1, \dots, j_k$ of $\widetilde{A}$. Then the relation $R_{1,\dots,k,j,\alpha}$ reflects the expansion of this determinant by minors along the αth row.

To show that the Grassmannian is determined by these relations it again suffices to consider the piece $\mathbb{C}^{\binom{n}{k}-1} = D^+(x_{1 \dots k})$ of $\mathbb{P}^{\binom{n}{k}-1}$. The relations $R_{1\dots k, j_1 \dots j_k, \alpha}$ with $j_1 \dots j_k \neq 1 \dots k$, will suffice to define the Grassmannian. First consider just those relations with $\alpha = 1$ and more than one of $j_1, \dots, j_k$ not belonging to the set $\{1, \dots, k\}$. There are $\binom{n}{k} - 1 - k(n-k)$ of these, and they may be indexed lexicographically by $j_1 \dots j_k$, as may be the coordinates $t_{j_1 \dots j_k}$ of $\mathbb{C}^{\binom{n}{k}-1}$. With respect to this ordering, consider the matrix

$$\frac{\partial \widetilde{R}_{1\dots k, j_1 \dots j_k, 1}}{\partial t_{i_1 \dots i_k}}$$

as both $j_1, \dots, j_k$ and $i_1, \dots, i_k$ run over the k-tuples with at least two members not in $\{1, \dots, k\}$. This will be a lower triangular square matrix, with nothing but positive or negative 1's and 2's along the diagonals (the 2's, as well as the off-diagonal entries, come from the columns gotten by differentiating with respect to $t_{i_1 \dots i_k}$ with $i_1 = 1$). From this it follows that the variety defined by the $R_{i,j,\alpha}$ is contained in a non-singular variety of dimension $k(n-k)$. Since this is the dimension of the Grassmannian one knows already that one irreducible component of the variety $V(R_{i,j,\alpha})$ is exactly the Grassmannian.

To complete the proof one must show that the variety $V(R_{i,j,\alpha})$ is irreducible; it will be enough to prove directly that $V(R_{i,j,\alpha}) \cap D^+(x_{1\dots k})$ coincides with the

Grassmannian. We have already seen that the imbedding

$$\mathbb{C}^{k(n-k)} \xrightarrow{\ p\ } \mathbb{C}^{\binom{n}{k}-1}$$

is essentially a graph, and that to each choice of coordinates $t_{1\ldots j\ldots k}$ (i^{th} place) with $i \le i \le k$, $k+1 \le j \le n$, there is exactly one choice of the remaining coordinates which will put the point in $p(\mathbb{C}^{k(n-k)})$. But the quadratic relations $\widetilde{R}_{1\ldots k, j_1\ldots j_k, \alpha}$, as α runs from one to k_j determine this choice: Given $j_1, \ldots, j_k$ with at least two of these greater than k, suppose that $j_\alpha > k$. Then the relation

$$t_{j_1\ldots j_k} = \Sigma_{q=1}^{k} (-1)^{q-1}\, t_{1\ldots j_q \ldots k}\; t_{j_1\ldots \hat{j}_q \alpha \ldots j_k}$$

(α^{th} place)

will express $t_{j_1\ldots j_k}$ in terms of $t_{i_1\ldots i_k}$ with at least one more of $i_1, \ldots, i_k$ in the set $\{1, \ldots, k\}$ than is the case with $j_1, \ldots, j_k$. One continues in this way until $t_{j_1\ldots j_k}$ is expressed in terms of $t_{1\ldots j\ldots k}$ (i^{th} place) by the quadratic relations.

It may seem that this last argument makes the earlier computation of the rank of a Jacobian matrix superflous. Yet this is not the case, since these two arguments together show that the ideal of polynomials vanishing on $p(\mathbb{C}^{k(n-k)})$ in $\mathbb{C}^{\binom{n}{k}-1}$ is actually generated by the $\widetilde{R}_{i,j,\alpha}$, and that the Grassmannian is determined by the quadratic relations in this strong sense.

Theorem M The Grassmannian manifold Grass (k, n) is a non-singular, irreducible, projective algebraic variety. The Plücker map

$$p : \mathrm{Grass}(k,n) \longrightarrow \mathbb{P}^{\binom{n}{k}-1}$$

is an imbedding, and the imbedded Grassmannian is the variety determined by the quadratic relations $R_{i,j,\alpha}$.

As we have already suggested, the Grassmannians are to higher dimensional bundles as the projective spaces are to line bundles : Grass (k, n) has on it a universal bundle of rank k. The bundle itself, which is denoted U_k, may be defined directly in terms of transition functions. There is the holomorphic map

$$\phi : St(k,n) \longrightarrow \mathrm{Grass}(k,n)$$

where St(k, n) is an open subset of the space of all kxn matrices, and Grass (k, n) is the quotient of the action of $GL(k,\mathbb{C})$ on St(k, n). For $1 \leq i_1 < \ldots < i_k \leq n$ there is the open set $\phi(U_{i_1 \ldots i_k}) = W_{i_1 \ldots i_k}$. There is a map

$$\eta_{i_1 \ldots i_k} : U_{i_1 \ldots i_k} \longrightarrow GL(k,\mathbb{C})$$

which takes a matrix to the matrix made of columns $i_1, \ldots, i_k$. For $A \in U_{i_1 \ldots i_k} \cap U_{j_1 \ldots j_k}$ set $\theta_{(\ell_1 \ldots \ell_k)(j_1 \ldots j_k)}(A) = \eta_{i_1 \ldots i_k}(A)^{-1} \eta_{j_1 \ldots j_k}(A) \in GL(k,\mathbb{C})$. Note that

$$\theta_{(i_1 \ldots i_k)(j_1 \ldots j_k)} = \theta_{(i_1 \ldots i_k)(\ell_1 \ldots \ell_k)} \theta_{(j_1 \ldots j_k)}$$

on $U_{i_1 \ldots i_k} \cap U_{\ell_1 \ldots \ell_k} \cap U_{j_1 \ldots j_k}$ and that these functions are invariant under the action of $GL(k,\mathbb{C})$. These define a k-bundle on $Grass(k,n)$, trivialized along the covering $\{W_{i_1\ldots i_k}\}$. In case $k=1$ the line bundle $O_{\mathbb{P}^{n-1}}(1)$ has been defined.

Now we shall define n global section of U_k over $Grass(k,n)$. For $\alpha \in \{1,\ldots,n\}$ and $1 \leq i_1 < \ldots < i_k \leq n$ we define $\phi_{\alpha,i_1\ldots i_k} : U_{i_1\ldots i_k} \longrightarrow \mathbb{C}^k$ by $\phi_{\alpha,i_1\ldots i_k} : A \longmapsto \eta_{i_1\ldots i_k}(A)^{-1}(\alpha^{th}$ column of $A)$. This will be invariant under the action of $GL(k,\mathbb{C})$ and therefore we get holomorphic maps

$$\phi_{\alpha,i_1\ldots i_k} : W_{i_1\ldots i_k} \longrightarrow \mathbb{C}^k$$

They are designed to satisfy the transition rules

$$\phi_{i_1\ldots i_k,\alpha} = \theta_{(i_1\ldots i_k)(j_1\ldots j_k)}\, \phi_{j_1\ldots j_k,\alpha}$$

and hence define global sections $\phi_1,\ldots,\phi_n$ of U_k. The sections $\phi_{i_1},\ldots,\phi_{i_k}$ will generate over $W_{i_1\ldots i_k}$.

Theorem N <u>On the Grassmannian manifold</u> $Grass(k,n)$ <u>there is a holomorphic vector bundle of rank</u> k, U_k, <u>with</u> n <u>global sections</u> $\phi_1,\ldots,\phi_n$ <u>which generate each fiber. These sections form a basis for</u> $H^0(Grass(k,n),U_k)$.

The last part of this theorem will be proven a little later.

A more conceptual description of the bundle U_k is in order here. It can be done in a few different ways. First recall that $St(k,n) \xrightarrow{\phi} Grass(k,n)$ is a $GL(k,\mathbb{C})$ bundle. There are biholomorphic maps $\psi_{i_1\ldots i_k}$, $\tilde{\psi}_{i_1\ldots i_k}$

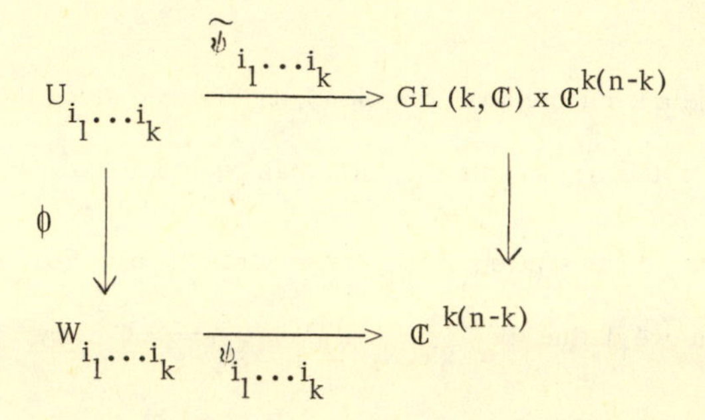

such that

$$\tilde{\psi}_{i_1 \ldots i_k} \circ \tilde{\psi}_{j_1 \ldots j_k}^{-1} \text{ is of the form}$$

$$(g, B) \longrightarrow \eta_{(i_1 \ldots i_k)(j_1 \ldots j_k)}(B) = \theta^{-1}_{(i_1 \ldots i_k)(j_1 \ldots j_k)}(B).$$

Thus $St(k, n) \longrightarrow Grass(k, n)$ is a principle bundle with group $GL(k, \mathbb{C})$ and the vector bundle associated to it is just the dual of U_k, U_k^*. $st(k, n)$ is the bundle of frames in U_k^*.

For another description, think of $Grass(k, n)$ as the k-spaces in $\mathbb{C}^n$, and consider the product

$$\mathbb{C}^n \times Grass(k, n)$$

There will be an algebraic subset V of $\mathbb{C}^n \times Grass(k, n)$ consisting of pairs (v, B) such that $v \in B$. The map $V \longrightarrow Grass(k, n)$ exhibits V as a $\mathbb{C}^k$-bundle which turns out to be U_k^*. One can see that V must be U_k^* because the bundle of frames may be identified with $St(k, n)$. This shows that U_k^* is a subbundle of a trivial bundle, and that U_k is the quotient of a trivial bundle. Denoting the locally free sheaf associated to U_k by $\Gamma(U_k)$, the map

$$O^n \xrightarrow{(\phi_1, \ldots, \phi_n)} \Gamma(U_k) \longrightarrow 0$$

exhibits U_k as a quotient.

Consider again the map ϕ^*: St (k, n) $\longrightarrow$ Grass (k, n). The $\mathbb{C}^k$ - bundle $\phi(U_k)$ is trivial, as one can see directly from the definition of the transition functions of U_k. By lifting the exhibition of U_k as a quotient, one gets a surjective map of trivial bundles

$$\text{St}(k, n) \times \mathbb{C}^n \longrightarrow \text{St}(k, n) \times \mathbb{C}^k$$

which is equivalent to a map

$$\psi : \text{St}(k, n) \longrightarrow \mathbb{C}^{kn}$$

considering $\mathbb{C}^{kn}$ as the space of $k \times n$ matrices. This map is of course the identity.

A holomorphic section of $\phi^*(U_k)$ over St (k, n) is just a holomorphic map f: St (k, n) $\longrightarrow \mathbb{C}^k$. Those sections of $\phi^*(U_k)$ which lift sections of U_k over Grass (k, n) are those f which satisfy

$$gf(A) = f(gA)$$

for all $g \in GL(k, \mathbb{C})$ and $A \in \text{St}(n, k)$ - this should be familiar from the computation of the global sections of $O_{\mathbb{P}^n}(1)$ on $\mathbb{P}^n$. Again one can use Hartog's extension theorem to show that such a lifted global section must extend to a holomorphic $f : \mathbb{C}^{kn} \longrightarrow \mathbb{C}^k$, satisfying the same relations under the action of $GL(k, \mathbb{C})$. Then a purely algebraic argument shows that a $\mathbb{C}$-basis for such maps is given by projecting onto the various columns of the $k \times n$ matrix, so that $\phi_1, \ldots, \phi_n$ are a basis for $H^0(\text{Grass}(k, n), U_k)$. (This completes the proof of theorem N).

Suppose given an analytic space X, with a holomorphic $\mathbb{C}^k$ - bundle $L \longrightarrow X$, generated by $\psi_1, \ldots, \psi_n \in H^0(X, L)$. Suppose that L is trivial along an open $U \subset x$. Then the ψ_i define a surjective bundle map

$$U \times \mathbb{C}^n \longrightarrow U \times \mathbb{C}^k$$

which is the same thing as a map

$$U \longrightarrow St(k, n)$$

The induced map $U \longrightarrow \text{Grass}(k, n)$ is independent of the choice of trivialization.

Theorem O _Given any analytic space_ X, _with holomorphic_ $\mathbb{C}^k$_-bundle_ $L \longrightarrow X$, _generated by_ $\psi_1, \ldots, \psi_n \in H^0(X, L)$ _there is a holomorphic map_

$$\psi : X \longrightarrow \text{Grass}(k, n)$$

such that $L \xrightarrow{\sim} \psi^*(U_k)$, $\psi_i = \psi^*(\phi_i)$. _Conversely, any holomorphic map_ $\psi : X \longrightarrow \text{Grass}(k,$ _induces a_ $\mathbb{C}^k$ - _bundle_ $\psi^*(U_k)$ _generated by the global sections_ $\psi^*(\phi_i)$, $i \in \{1, \ldots, n\}$.

This is exactly as in the case of projective space.

It follows from this theorem, as in the case of projective space, that the group of holomorphic automorphisms of $\text{Grass}(k, n)$ may be identified with $PGL(n, \mathbb{C})$. We already saw that this group of automorphisms operates transitively.

One could approach the Grassmannian from an algebraic point of view and develop a parallel theory for algebraic vector bundles on algebraic varieties. From the fact that the universal bundle U_k is algebraic, and from Chow's theorem, we get another proof of the GAGA result that holomorphic vector bundle theory and algebraic bundle theory will be the same on a projective variety. For given a bundle L on the projective variety $X \longrightarrow \mathbb{P}^n$, one can find d such that $L(d) = L \otimes_{O_X} O_{\mathbb{P}^n}(d)$ will be generated by its global sections and therefore induced by a holomorphic (whence algebraic) map to the Grassmannian.

To investigate the structure of the Grassmann manifolds the Schubert varieties will be introduced here. Again think of Grass (k,n) as the manifold of k-spaces in C^n, and St (k,n) as the space of k-frames in C^n. Fix a filtration $0 \subset L_1 \subset L_2 \subset \ldots \subset L_n = C^n$ of subspaces of C^n, with L_i a space of dimension i. For any k-tuple of integers $(a_1, a_2, \ldots, a_k)$ with $0 \le a_1 \le a_2 \le \ldots \le a_k \le n-k$, the Schubert variety of symbol $(a_1, \ldots, a_k)$ is defined to be the set of all $x \in$ Grass (k,n) such that

$$\dim x \cap L_{a_i+i} \ge i$$

for $i \in \{1, \ldots, k\}$. Each $(a_1, \ldots, a_k)$ will be an algebraic subset of the Grassmannian, as one might see by considering the sets $\phi^{-1}((a_1, \ldots, a_k)) \in$ St (k,n). Supposing that L_i is defined by $z_n = z_{n-1} = \ldots = z_{n-i} = 0$ and thinking of $A \in$ St (k,n) as a $k \times n$ matrix, $\phi^{-1}((a_1, \ldots, a_k))$ is defined by the conditions

$$\text{rank} \begin{pmatrix} \cdots\cdots\cdots\cdots \\ \cdots\cdots\cdots\cdots \\ 0 \quad 1 \quad 0 \ldots\ldots 0 \\ 1 \quad 0 \quad 0 \ldots\ldots 0 \\ A \end{pmatrix} \begin{matrix} a_i+i \text{ rows} \\ \\ \le k + a_i \end{matrix}$$

for $i \in \{1, \ldots, k\}$, which are obviously algebraic conditions, and invariant under the action of $GL(k, \mathbb{C})$.

Little attention has been paid to the particular choice of filtration because $GL(n, \mathbb{C})$ acts transitively on filtrations in its action on the Grassmannian. One may as well assume that L_i is given by $z_n = \ldots = z_{n-i} = 0$.

The Schubert varieties are to the Grassmannians as the hyperplanes are to the projective spaces. We shall be particularly concerned with the cohomology classes of the Schubert varieties : As we shall see later in this chapter, to each analytic subvariety X of pure codimension k on the complex manifold M there is associated a cohomology class $[X] \in H^{2k}(M, \mathbb{Z})$. This generalizes what we have done for divisors and has similar functorial properties.

Theorem P Each symbol $(a_1, \ldots, a_k)$ defines an algebraic subvariety of the Grassmannian, of pure dimension $a_1 + \ldots + a_k$. The cohomology classes of these Schubert varieties form a basis for the integral cohomology of the Grassmannian, which thus has non-zero cohomology only in even dimensions and is free of torsion.

The statement about the dimension of the Schubert variety with symbol $(a_1, \ldots, a_k)$ be proved by considering the varieties $\phi^{-1}((a_1, \ldots, a_k)) \in St(k, n)$. We already have a description of the equations defining these varieties, and an analysis of these shows that the codimension of $\phi^{-1}((a_1, \ldots, a_k))$ in $St(k, n)$ is $(n-k)k - (a_1 + \ldots + a_k)$ and because $\phi : St(k, n) \longrightarrow Grass(k, n)$ is a smooth map this determines the dimension of $(a_1, \ldots, a_k)$. A different proof appears in Chern [8]. This book also contains a fuller discussion of the rest of the theorem. A topological discussion of the Grass -mannians appear in Bott [36].

A Schubert variety $(a_1, \ldots, a_k)$ with $a_i > a_{i-1}$ $(a_o = 0)$ properly contains the Schubert variety $(a_1, \ldots, a_{i-1}, a_i - 1, a_{i+1}, \ldots, a_k)$. Setting

$$(a_1, \ldots, a_k)^* = (a_1, \ldots, a_k) - \sum_{a_i > a_{i-1}} (a_1, \ldots, a_{i-1}, a_i - 1, \ldots, a_k)$$

it is proven in the book of Chern that $(a_1, \ldots, a_k)^*$ is a complex manifold which is topologically a cell of real dimension $2(a_1 + \ldots + a_k)$. This gives a cell decomposition of the Grassmannian, as well as information about the singularities of the Schubert varieties.

The Schubert varieties

$$c_i(U_k) = (\underbrace{n-k-1, \ldots, n-k-1,}_{i \text{ places}} n-k, \ldots, n-k)$$

for $i = 0, \ldots, k$, are singled out for special attention. $c_i(U_k)$, or more properly the cohomology class $c_i(U_k)$ in $H^{2i}(\text{Grass}(k, n), \mathbb{Z})$, is called the universal i^{th} Chern class. This terminology will be explained later.

These Schubert varieties have a nice interpretation. $\phi^{-1}((a_1, \ldots, a_k))$ is described by the conditions

$$\text{rank} \begin{pmatrix} I_{i+a_i} + O \\ \\ A \end{pmatrix} \leq k + a_i, \quad i \in \{1, \ldots, k\}$$

or equivalently

rank (matrix of last $n - a_i - i$ columns of A) $\leq k - i$, $i \in \{1, \ldots, k\}$.

Recall the previous notation that $\phi_1, \ldots, \phi_n$ are global sections of U_k over $\text{Grass}(k, n)$.

From the last description of the Schubert varieties it follows that

$c_i(U_k) = \{$points where the sections $\phi_n, \phi_{n-1}, \ldots, \phi_{n-(k-i)}$

are linearly dependent$\}$.

In particular, $c_k(U_k) = \{$points where $\phi_n = 0\}$

$$c_0(U_k) = \text{Grass}(k,n)$$

$c_1(U_k) = \{$points where $\phi_n, \ldots, \phi_{n-k+1}$

are linearly dependent$\}$

More properly, one speaks of the cohomology classes defined by these conditions.

In case $k = 1$, U_k is $O_{\mathbb{P}^{n-1}}(1)$ and $c_1(U_k)$ is represented by a hyperplane. More generally, there is the Plücker imbedding

$$p : \text{Grass}(k,n) \longrightarrow \mathbb{P}^{\binom{n}{k}-1}$$

Now $O_{\mathbb{P}^{\binom{n}{k}-1}}\Big|_{\text{Grass}(k,n)}$ is $\bigwedge^k U_k$, as may be seen from comparison of the transition functions. Also $\phi_n \ldots \phi_{n-k+1}$ defines a global section of $\bigwedge^k U_k$, indeed one induced from a section of $O_{\mathbb{P}^{\binom{n}{k}-1}}(1)$ over $\mathbb{P}^{\binom{n}{k}-1}$, so $c_1(U_k)$ is always the intersection of Grass (k,n) with a hyperplane. It should be noted that $\phi_n \ldots \phi_{n-k+1}$ is usually not a generic section of $\bigwedge^k U_k$. In fact, the Schubert varieties representing $c_1(U_k)$ always have singularities for $k > 1$, while the intersection of the Grassmannian with a generic hyperplane is non-singular. The Schubert variety representing $c_k(U_k)$ deserves special consideration. It has the symbol

$$(n - k - 1, \ldots, n - k - 1)$$

and can be described as the set of k-spaces in $\mathbb{C}^n$ which are contained in the subspace defined by $z_n = 0$. This suggests that the Schubert variety may be identified with Grass (k, n-1). In fact the imbedding

$$St(k, n-1) \longrightarrow St(k, n)$$

$$\text{by } A \longrightarrow (A \vdots\ )$$

induces an isomorphism of Grass (k, n-1) with this Schubert variety. In particular, $c_k(U_k)$ is non-singular. In general $c_i(U_k)$ has singularities for $i \neq 0, k$ and the singularities of $c_i(U_k)$ are contained in $c_{i+1}(U_k)$ - these results are contained in Kleiman [37].

The simplest Grassmannian which is not a projective space is Grass(2, 4), which may be thought of as the lines in $\mathbb{P}^3$. It is a 4-fold, and the Plücker imbedding

$$p : \text{Grass}(2, 4) \longrightarrow \mathbb{P}^5$$

gives it as the hypersurface defined by

$$x_{12}x_{34} - x_{32}x_{14} + x_{42}x_{13} = 0$$

or

$$x_{12}x_{34} + x_{23}x_{14} - x_{24}x_{13} = 0$$

using homogeneous coordinates $x_{12}, x_{13}, x_{14}, x_{23}, x_{24}, x_{34}$ in $\mathbb{P}^5$.

The Schubert varieties have symbols

0 - dimensional { (0, 0)

1 - dimensional { (0, 1)

2 - dimensional { (0, 2)
(1, 1)

3 - dimensional { (1, 2)

4 - dimensional { (2, 2)

so the Betti numbers are $b_0 = b = 1$, $b_2 = b_6 = 1$ $b_4 = 2$, $b_{odd} = 0$. $c_1(U_2)$ is represented by the variety $V^+(x_{34})$ on the Grassmannian and will have one singularity, at the point (1, 0, ..., 0). $c_2(U_2)$ will be isomorphic to $\mathbb{P}^2$ and defined by $V^+(x_{14}, x_{24}, x_{34})$. See Šafarevič [35] for a discussion of how one uses Grass (2, 4) to determine which surfaces in $\mathbb{P}^3$ have lines on them, and how many.

Given a complex manifold M, with a holomorphic $\mathbb{C}^k$-bundle L induced by a map

$$\phi : M \longrightarrow \text{Grass}(k, n),$$

the cohomology classes $\phi^*(c_i(U_k)) \in H^{2i}(M, \mathbb{Z})$ are called the Chern classes of L. It follows from this definition that the Chern classes are represented by analytic subvarieties. Although it is not clear from this definition, the Chern classes depend only on the isomorphism class the bundle L, not on the map ϕ.

Two main theorems to be discussed later are:

Theorem Q Let the complex manifold M be either an affine algebraic variety or a projective variety. Then the subring of the cohomology ring $H^{\bullet}(M, \mathbb{Q})$ generated by analytic cohomology classes (that is, cohomology classes of analytic subvarieties) is the subring generated by the Chern classes of holomorphic vector bundles.

Theorem R Let the complex manifold M be affine algebraic. Then the ring $H^{\text{even}}(M, \mathbb{Q})$ is generated by the Chern classes of holomorphic vector bundles.

Both these theorems are some distance away. They require all we have done so far, together with the Bott periodicity theorem and a deep theorem of Grauert, both to be discussed in the next chapters.

Chapter Five

Section Three — Chern classes and curvature

In this section we shall discuss vector bundles and their characteristic cohomology classes from the point of view of topology and differential geometry. We shall connect the topological and differential results and then focus on the special case of holomorphic bundles on complex manifolds.

For the rest of our notes all our topological spaces will be connected, locally compact, countably compact CW complexes of finite dimension, unless we make special mention to the contrary. Also, our spaces will almost always have the homotopy types of finite simplicial complexes (see Spanier [29] for our topological terminology).

The investigation of the characteristic classes of complex vector bundles will make use of the splitting principle: If $E \longrightarrow X$ is a continuous complex vector bundle, there is canonically constructed another space Y with a map $\pi : Y \longrightarrow X$ such that

(1) $\pi^* : H^{\cdot}(X, \mathbb{Z}) \longrightarrow H^{\cdot}(Y, \mathbb{Z})$ is an injection

(2) $\pi^*(E)$ is a sum of line bundles.

The construction of $\pi: Y \longrightarrow X$ is inductive. We first construct $\pi_1 : Y_1 \longrightarrow X$ such that π_1^* is injective on integral cohomology and $\pi_1^*(E)$ splits off a line bundle as a direct summand, $\pi_1^*(E) = L \oplus E'$ where L is a line bundle. Then we do the same thing with $E' \longrightarrow Y_1$.

$\pi_1 : Y_1 \longrightarrow X$ is constructed as a projective bundle over X. If $E \longrightarrow X$ has fiber $\mathbb{C}^d$, $\pi_1 : Y_1 \longrightarrow X$ is the associated bundle of projective spaces, with fiber $\mathbb{P}^{d-1}$, denoted $\mathbb{P}(E)$. A point of $\mathbb{P}(E)$ corresponds to a line through the origin of a fiber of E. $\mathbb{P}(E)$ may be constructed by letting $\mathbb{C}^x$ act on E - {zero section} and then taking the quotient.

In this construction $\mathbb{P}(E)$ is provided with a tautological line bundle L, such that L restricted to any fiber is $O_{\mathbb{P}^{d-1}}(-1)$. L is constructed by pulling back E to $\mathbb{P}(E)$ then assigning to each point of $\mathbb{P}(E)$ the line through the fiber which it represents. L is naturally a sub line bundle of $\pi_1^*(E)$.

There is then an exact sequence of bundles on $\mathbb{P}(E)$

$$0 \longrightarrow L \longrightarrow \pi_1^*(E) \longrightarrow \pi_1^*(E)/L \longrightarrow 0$$

and we want to show that this sequence splits, so $\pi_1^*(E) \xrightarrow{\sim} L \oplus \pi_1^*(E)/L$. We will use the fact that $\mathbb{P}(E)$ is a paracompact space, so $\pi^*(E)$ admits a Hermitian metric, as was discussed in Section one of this chapter. Then $\pi_1^*(E)/L$ is identified with the orthogonal complement bundle to L, and $\pi_1^*(E)$ is the direct sum. (This argument shows that, on good spaces, exact sequences of bundles always split.)

THEOREM S: <u>The integral cohomology of</u> $\mathbb{P}(E)$ <u>is a free module over the integral cohomology of</u> X, <u>with basis</u> $1, c_1(L), [c_1(L)]^2, \ldots, [c_1(L)]^{d-1}$.

The proof of this theorem comes from the Leray spectral sequence of the fibration $\mathbb{P}(E) \longrightarrow X$. A discussion appears in Spanier [29].

It follows from this theorem that there is a relation

$$(-1)^d [c_1(L)]^d + (-1)^{d-1} \; c_1(E)[c_1(L)]^{d-1} + \ldots + c_d(E) = 0$$

where the $c_i(E)$ are uniquely determined, $c_i(E) \in H^{2i}(X, \mathbf{Z})$. The c_i so defined is called the ith Chern class of the vector bundle E_1 for $i \in \{1, \ldots, d\}$. The definition is extended by setting $c_0(L) = 1$. The $c(E)$ is used to denote $c_0(E) + c_1(E) + \ldots + c_j(E) \in H^{\cdot}(X, \mathbf{Z})$ and is called the total Chern class of E. Note that in case E is a line bundle this definition of $c_1(E)$ agrees with the previous one.

THEOREM T: The Chern classes have the properties

(1) $c(f^*E) = f^*(c(E))$ for $f: Y \longrightarrow X$, $E \longrightarrow X$ on a bundle

(2) $c(E \oplus W) = c(E)\, c(W)$ for $E \longrightarrow X$, $W \longrightarrow X$ bundles.

Part (1) of this theorem follows from the functoriality of construction of $\mathbb{P}(E)$. To get part (2), we first prove: If $E = L_1 \oplus L_2 \oplus \ldots \oplus L_d$ is a direct sum of line bundles then $c(E) = \Pi(1 + c_1(L_i))$. From the definition of the Chern classes, this is equivalent to showing $\Pi(c_1(L_i) - c_1(L)) = 0$ on $\mathbb{P}(E)$.

Now L is a sub line bundle of $\Pi^*(E) = \Pi_1^*(L_1) \oplus \ldots \oplus \Pi_1^*(L_d)$, so $\Pi_1^*(L_1) \otimes L^{-1} \oplus \ldots \oplus \Pi_1^*(L_d) \otimes L^{-1}$ contains a trivial sub line bundle. This means that $\Pi_1^*(E) \otimes L^{-1}$ has a non-vanishing global section s. We can write $s = \Sigma\, s_i$, where s_i is a global section of $\Pi_1^*(L_i) \otimes L^{-1}$. Let U_i be the open set where s_i does not vanish; then $\mathbb{P}(E)$ is the union of the U_i. Now $c_1(L_i) - c_1(L)$ is zero in $H^2(U_i, \mathbf{Z})$, and $(c_1(L_i) - c_1(L))(c_1(L_j) - c_1(L))$ is zero in

$H^4(U_i \cup U_j, \mathbb{Z})$. Continuing in this way, $\Pi(c_1(L_i) - c_1(L))$ is zero in $H^{2d}(\mathbb{P}(E), \mathbb{Z})$.

Now part (2) will follow from this and the splitting principle.

Before moving on, we should mention that the space $Y \longrightarrow X$, on which the pull-back of E splits into a sum of line bundles, is called the flag space of $E \longrightarrow X$. A fiber of $Y \longrightarrow X$ consists of all possible filtrations $0 \subset E_{1,x} \subset E_{2,x} \subset \ldots \subset E_{d,x}$ with one-dimensional successive quotients, of the fiber of E over a point $x \in X$. This follows from the inductive construction of Y.

THEOREM U: On the Grassmannian Grass (k, n) the Chern classes of the universal bundle are the $c_i(U_k)$ as previously defined.

The proof is deferred to the end of Section Four of this chapter.

Now we shall restrict our attention to differentiable manifolds and differentiable vector bundles, seeing how to represent Chern classes by differential forms

Let M be a differentiable manifold, $E \longrightarrow M$ a differentiable complex vector bundle of rank d. On M there is the sheaf of sections of E, which we shall also denote by E, and the sheaf of complex 1-forms, denoted by $T(M)^*$. A connection on E is a map

$$D : E \longrightarrow T(M)^* \otimes_{\mathbb{C}} E$$

satisfying

$$D(fe) = df \otimes e + fDe$$

for a function f and a section e of E.

A connection on a bundle provides a way to differentiate sections of that bundle: If v is a tangent vector at some point, and e is a section of E defined in a neighborhood of that point, then the derivative of e in the direction v is

$$D_v(e) = De(v) ,$$

an element of the fiber of E over our point.

Continuing in this vein, suppose that $e_1, \ldots, e_d$ is a frame of E over an open set (that is, which gives a basis in every fiber). Then a section over this open can be expressed as $e = f_1 e_1 + \ldots + f_d e_d$ with smooth functions f_i, and the derivative of e with respect to v will be

$$D_v e = v(f_1)e_1 + \ldots + v(f_d)e_d + f_1 D_v e_1 + \ldots + f_d D_v e_d .$$

Thus to differentiate a section we pick a frame, differentiate componentwise, then add on a correction term depending on the frame.

We can write

$$De_i = \Sigma\, \omega_{ij} e_j ,$$

the ω_{ij} being 1-forms. Then (ω_{ij}) is called the <u>connection matrix</u> with respect to the frame $e_{ij}, \ldots, e_d$.

Given a connection we define maps

$$D^{(i)} e \otimes \alpha = (-1)^i De \wedge \alpha$$

$$D^{(i)} : \wedge_{\mathbb{C}} T(M)^* \otimes E \longrightarrow \wedge^{i+1}_{\mathbb{C}} T(M)^* \otimes E$$

by

$$D^{(i)}\alpha \otimes e = d\alpha \otimes \alpha + (-1)^{i} \alpha \wedge De \quad .$$

The map $D^{(i)}D = D^2 : E \longrightarrow \wedge^2 T(M)^* \otimes E$ satisfies

$$D^2(fe) = D^{(i)}(df \otimes e + f\,De)$$

$$= -df \wedge De + df \wedge De + fD^2e$$

$$= f\,D^2e$$

and is threrfore a linear map of bundles. This map is called the curvature of the connection.

In terms of a frame $e_1, \ldots, e_d$ we have

$$D^2 e_i = \Sigma\, \Omega_{ij}\, e_j \quad .$$

The matrix of 2-forms (Ω_{ij}) is called the curvature form matrix with respect to the frame $e_1, \ldots, e_d$. We can compute it in terms of the connection form:

$$D^2 e_i = D^{(i)}(\Sigma\, \omega_{ij}\, e_j)$$

$$= \Sigma\, d\omega_{ij} e_j - \Sigma\, \omega_{ik} \wedge \omega_{kj}\, e_j$$

so

$$\Omega = d\omega - \omega \wedge \omega \quad .$$

The curvature can also be expressed in terms of the Lie derivative (this is because the exterior derivative of forms can be defined in terms of the Lie derivative).

If n, v are vector fields over an open set and e is a section of E over that set then

$$D_n(D_v e) - D_v(D_n e) - D_{[n, v]} e$$

is equal to $D^2 e(n, v)$. We give a proof by computation in local frames:

$$D_n(D_v e_i) - D_v(D_n e_i) - D_{[n, v]} e_i$$

$$= D_n(\Sigma \omega_{ij}(v) e_j) - D_v(\Sigma \omega_{ij}(n) e_j) - \Sigma \omega_{ij}([n, v]) e_{ij} \quad .$$

Now the relation between exterior differentiation and the Lie bracket is

$$d_w(n, v) = n\, w(v) - v w(n) - w([n, v])$$

for a 1-form w . Using this in our last expression yields

$$D_n(D_v e_i) - D_v(D_n e_i) - D_{[n, v]} e_i$$

$$= \Sigma\, d\omega_{ij}(n, v)\, e_j - \Sigma\Sigma\, \omega_{ik} \wedge \omega_{kj}(n, v)\, e_j$$

which is the curvature. (For a discussion of the relation between Lie differentiation and exterior differentiation, see Hicks [16], and the references given there.)

Connections can also be defined in other ways, some of which we shall use here. For instance, let $E \longrightarrow M$ be a complex vector bundle (differentiable) over a differentiable manifold M . The total space E is also a differentiable manifold with sheaf of tangent vectors $T(E)$. We denote by $\widetilde{V}(E)$ the subsheaf of tangent vectors which are linear along the fibers. To see what this means, note that if E is trivial, $E = \mathbb{C}^d \times M$. Then $T_{\mathbb{C}}(E) = \mathbb{C}^d \times T_{\mathbb{C}}(M)$. A vector field $E \longrightarrow T(E)$ is called linear if the induced map $\mathbb{C}^d \longrightarrow \mathbb{C}^d$ is linear. This definition is invariant under the trivialization and in this way the sheaf $\widetilde{V}(E)$ is defined. The restriction of $\widetilde{V}(E)$ to the zero section is a sheaf on M, denoted $V(E)$. The map $T_{\mathbb{C}}(E) \xrightarrow[\text{o-section}]{} T_{\mathbb{C}}(M)$ induces a map

$$V(E) \longrightarrow T_{\mathbb{C}}(M)$$

which is surjective. The kernel may be identified with $\underline{\mathrm{Hom}}(E,E)$ and there is an exact sequence

$$0 \longrightarrow \underline{\mathrm{Hom}}(E,E) \xrightarrow{i} V(E) \xrightarrow{\psi} T(M) \longrightarrow 0$$

$\widetilde{V}(E)$, as a subsheaf of $T_{\mathbb{C}}(E)$, inherits the structure of Lie algebra, as does $V(E)$. $T_{\mathbb{C}}(M)$ is also a sheaf of Lie algebras.

In this framework, a connection in E is a map $\psi\colon T_{\mathbb{C}}(M) \longrightarrow V(E)$ which splits the exact sequence. The curvature of the connection is an alternating map from $T_{\mathbb{C}}(M) \times T_{\mathbb{C}}(M)$ to $\underline{\mathrm{Hom}}(E,E)$, defined by

$$i(R(t,n)) = \psi([t,n]) - [\psi(t), \psi(n)]$$

which is an element of $\underline{\mathrm{Hom}}(E,E)$.

This point of view is useful to show the existence of connections for, as we have seen, an exact sequence of vector bundles on a paracompact differentiable manifold always admits a splitting.

To show that this definition of connection is equivalent to the previous one, we begin by showing how a connection map

$$D : E \longrightarrow E \otimes_{\mathbb{C}} (T(M)^*)$$

gives a splitting. If $e_1, \ldots, e_d$ is a frame for E over an open set, and $D(e_i) = \Sigma\, \omega_{ij}\, e_j$, then the map $\psi : T(M) \longrightarrow V(E)$ is given by

$$\psi : t \longrightarrow (\omega_{ij}(t), t)$$

under the local isomorphism

$$V(E) \xrightarrow{\sim} \underline{\mathrm{Hom}}(E,E) \oplus_{\mathbb{C}} T(E) \quad .$$

Conversely, given a splitting φ and choosing a basis $e_1, \ldots, e_n$ one gets a matrix of 1-forms w_{ij}. Some computations in local coordinates show that these definitions are the same, and that the definition of curvature is consistent.

After choosing a frame $\alpha = \{e_1, \ldots, e_n\}$ the curvature is given by a matrix of 2-forms Ω_α. If the frame is changed to $\beta = \{f_1, \ldots, f_n\}$ with $(f_1, \ldots, f_n) = g_{\beta\alpha}(e_1, \ldots, e_n)$ then the change of the curvature matrix is given by

$$\Omega_\beta = g_{\beta\alpha}\, \Omega_\alpha\, g_{\beta\alpha}^{-1} \quad .$$

This shows that $\det(\Omega_\beta) = \det(\Omega_\alpha)$ is a well-defined differential form, as is $\mathrm{tr}(\Omega_\beta) = \mathrm{tr}(\Omega_\alpha)$. By locally expressing the curvature in frames, we get an expression

$$\det (tI_d + \frac{1}{2\pi\sqrt{-1}}\Omega) = t^d + c_1(\Omega)t^{d-1} + \ldots + c_d(\Omega)$$

where $c_i(\Omega)$ is a globally defined complex-valued differential form of degree $2i$.

THEOREM V: The differential forms $c_i(\Omega)$ are closed. If Ω^1 is the curvature induced by another connection in E, then $c_i(\Omega^1) - c_i(\Omega)$ is exact. The cohomology class defined by $c_i(\Omega)$ is the ith Chern class of E_i in $H^{2i}(M, \mathbb{C})$.

We shall give a complete proof of this theorem only in the case where M is a compact differentiable manifold. A proof in the general case appears in Kobayshi-Nomizu [18].

Proceeding with our proof, consider the space of $d \times d$ complex matrices, $gl(n, \mathbb{C})$. A polynomial function $p : gl(d, \mathbb{C}) \longrightarrow \mathbb{C}$ is called invariant in case $p(gAg^{-1}) = p(A)$ for $A \in gl(d, \mathbb{C})$, $g \in GL(d, \mathbb{C})$. The invariant polynomials form a ring. Particular examples of invariant polynomials are the functions $p_i(A)$ defined by

$$\det (tI_j + A) = t^n + p_1(A)\, t^{n-1} + \ldots + p_n(A)$$

and the $s_i(A)$ defined by

$$\operatorname{trace} (A^i) = s_i(A) \quad .$$

The ring of invariant polynomials is generated over $\mathbb{C}$ by either the $p_i(A)$ or the $s_i(A)$, $i \in \{1, \ldots, n\}$.

If Ω is the local curvature matrix of a connection on a vector bundle E of rank d, and p is an invariant polynomial, then $p(\Omega)$ is a globally defined differential form. Letting I denote the ring of invariant polynomials, we get a homomorphism

$$W : I \longrightarrow \mathrm{Hom}(\wedge T(M), \mathbb{C})$$

called the Weil homomorphism by

$$W : p \longrightarrow p\left(\frac{-1}{2\pi\sqrt{-1}}\,\Omega\right)$$

The image of this homomorphism is contained in the ring of closed 2-forms. For it suffices to prove that one always has

$$d\, s_i\left(\frac{1}{2\pi\sqrt{-1}}\,\Omega\right) = d\,\mathrm{trace}\left(\left(\frac{1}{2\pi\sqrt{-1}}\,\Omega\right)^i\right) = 0 \quad .$$

Now

$$d\,\mathrm{trace}\left(\left(\frac{1}{2\pi\sqrt{-1}}\,\Omega\right)^i\right) = \left(\frac{1}{2\pi\sqrt{-1}}\right) \Sigma_{R=1}^{i}\ \mathrm{Trace}\ (\Omega \ldots \underset{k^{th}\text{ place}}{d\Omega} \ldots \Omega) \quad .$$

Recall that if the connection matrix is expressed in local coordinates as (ω_{ij}) then the curvature matrix is expressed in the same local coordinates as

$$\Omega_{ij} = d\omega_{ij} - \Sigma_{R=1}^{n} \omega_{ik} \wedge \omega_{kj}$$

so that $d\Omega = -d\omega \wedge w + \omega \wedge dw$.

Now if $g \in GL(n, \mathbb{C})$ and $A_1, \ldots, A_i \in gl\,(n, \mathbb{C})$ then $\mathrm{Trace}\,(A_1, \ldots, gA_k - A_k g\,, \ldots, A_i) = 0$ as we see by considering terms linear in g in the identity

$$\mathrm{Trace}\,(A_1, \ldots, (I+g)A_k(I - g + g^2 - g^3 + \ldots), \ldots A_i) = \mathrm{Trace}\,(A_1, \ldots, A_i)$$

Extending this to the case where the A_i are matrices of 2-forms and the g is a matrix of 1-forms, and using the expression for $d\Omega$ we get the desired result.

The Weil homomorphism then reduces to a map into the complex cohomology algebra

$$W : I \longrightarrow H^{\cdot}(M, \mathbb{C})$$

It remains to show that this map is independent of the connection in the bundle.

Let then w_0, w_1 be two connection matrices in local coordinates. For each $t \in [0, 1]$ a connection is defined by $(1-t)w_0 + tw_1 = w_t$, with curvature Ω_t. Setting $\eta = w_1 - w_0$, one gets

$$\Omega_t = \Omega_0 - t(-d\eta + w_0 \wedge \eta + \eta \wedge w_0) - t^2 (\eta \wedge \eta) \qquad .$$

For $i \in \{1, \dots, n\}$, set

$$\eta_t^i = \operatorname{Trace}(\eta\, \underbrace{\Omega_t, \dots, \Omega_t}_{i-1 \text{ times}})$$

$d\eta_t^i = \operatorname{Trace}(d\eta, \Omega_t, \dots, \Omega_t)$, by a computation similar to that made before. Also,

$$\begin{aligned} \frac{d}{dt} \operatorname{Trace}(\Omega_t, \dots, \Omega_t) = {} & i \operatorname{Trace}(d\eta\, \Omega_t, \dots, \Omega_t) \\ & -i \operatorname{Trace}(w_0 \wedge \eta + \eta \wedge w_0, \Omega_t, \dots, \Omega_t) \\ & -i2t \operatorname{Trace}(\eta \wedge \eta, \Omega_t, \dots, \Omega_t) \\ = {} & i \operatorname{Trace}(d\eta, \Omega_t, \dots, \Omega_t) \qquad . \end{aligned}$$

Therefore

$$\text{Trace}(\Omega_1, \ldots, \Omega_1) - \text{Trace}(\Omega_0, \ldots, \Omega_0) = d\int_0^1 i\,\text{Trace}(d\tau, \Omega_t, \ldots, \Omega_t)\,dt$$

which shows what we wanted.

Before showing that the appropriate differential forms represent the Chern classes, some remarks about connections in complex vector bundles are in order. As we have mentioned before, if $E \longrightarrow M$ is a complex vector bundle on a differentiable manifold, then E always admits an hermitian metric. Specializing to the case of a holomorphic vector bundle on a complex manifold, provided with an hermitian metric, there is always a unique connection D in E such that

(1) $$d\langle e, f\rangle = \langle De, f\rangle + \langle e, Df\rangle$$

for differentiable sections e, f of E on an open set.

(2) The connection matrix of D is represented by forms of type $(1, 0)$.

To show that such a connection exists, suppose that E is given by the transition functions $\{g_{ij}\}$ and the hermitian metric is given by $\{h_i\}$. The connection matrix of D may be taken as $w_i = h_i^{-1}\partial h_i$, which is of type $(1, 0)$ and transforms properly.

To show uniqueness, note that a connection which preserves the hermitian structure must satisfy, in its local connection matrix

$$dh_i = w_i h_i + h_i^t \overline{w_i}$$

nad if w_i is of type $(1, 0)$ it must be $h_i^{-1}\partial h_i$. (The connection matrix will be of type $(1, 0)$ only with respect to holomorphic frames.)

The curvature of such a connection is given by

$$\Omega_i = d(h_i^{-1} \partial h_i) + h_i^{-1} \partial h \wedge h_i^{-1} \partial h$$

$$= \partial(h_i^{-1}) \wedge \partial h_i - h_i^{-1} \partial h \wedge h_i^{-1} \partial h$$

$$+ \bar{\partial}(h_i^{-1}) \wedge \partial h$$

$$+ h_i^{-1} \bar{\partial} \partial h_i$$

$$= h_i^{-1} \bar{\partial} \partial h - h_i^{-1} \bar{\partial} h \wedge h^{-1} \partial h$$

a matrix of forms of type $(1, 1)$.

In particular, if a holomorphic line bundle L, with transition functions $\{\alpha_{ij}\}$ is given, then a hermitian metric consists in a map $a_i : v_i \longrightarrow \mathbb{R}^+$ such that

$$|\alpha_{ij}|^2 = a_j / a_i$$

and the curvature of the associated connection is the 2-form

$$\bar{\partial} \partial \log a_i \quad .$$

Tracing through the co boundary map $\delta : H^1(M, O^*) \longrightarrow H^2(M, \mathbb{Z})$, $c_1(L)$ is represented by

$$\beta_{ijk} = \frac{1}{2\pi i} (\log \alpha_{jk} - \log \alpha_{ik} + \log \alpha_{ij})$$

in $c^2(M, \mathbb{Z})$. To represent this under the deRham isomorphism, we need 1-forms δ_k such that

$$\frac{1}{2\pi i} \, d \log \alpha_{jk} = \gamma_k - \gamma_j \qquad .$$

Then $d\gamma_i$ will represent $c_1(L)$. Since

$$\log \alpha_{jk} + \log \alpha_{jk} = \log a_k - \log a_j$$

$$d \log \alpha_{jk} = \partial \log \alpha_{jk} = \partial \log a_k - \partial \log a_j$$

so we can take $\gamma_k = \frac{1}{2\pi\sqrt{-1}} \partial \log a_k$, and

$$c_1(L) = \frac{1}{2\pi\sqrt{-1}} \bar{\partial}\partial \log a_k \qquad .$$

This shows how to represent the first Chern class of holomorphic line bundles on a complex manifold in terms of the curvature.

To extend this, observe that if $f : N \longrightarrow M$ is a differentiable map of differentiable manifolds, and $E \longrightarrow M$ is a complex vector bundle with connection D , then f^*E inherits a natural connection f^*D . Furthermore, the invariant differential forms associated to f^*D are pulled back from those associated to D on M .

If D_1 , D_2 are connections on bundles, E , E_2 , there is a natural connection $D_1 \oplus D_2$ on $E_1 \oplus E_2$. The curvature of this connection is given by the matrix

$$\begin{pmatrix} \Omega_1 & 0 \\ 0 & \Omega_2 \end{pmatrix}$$

so that

$$p_i\left(\frac{1}{2\pi\sqrt{-1}} \begin{pmatrix} \Omega_1 & 0 \\ 0 & \Omega_2 \end{pmatrix}\right) = \sum_{j+k=1} p_j\left(\frac{1}{2\pi\sqrt{-1}} \Omega_1\right) p_k\left(\frac{1}{2\pi\sqrt{-1}} \Omega_2\right)$$

Now by using a flag manifold we get the theorem for the Chern classes of holomorphic bundles on the complex manifolds- we must observe that the flag manifold construction will stay in the holomorphic category if we start with holomorphic bundles. In particular, we get the result for the universal bundle on the Grassmannian.

Now any differentiable complex vector bundle on a compact differentiable manifold is induced by a differentiable map to a Grassmannian: Any bundle on a compact manifold is generated by finitely many global sections, which induce a map to the Grassmannian just as in the holomorphic case. Thus we get the result for compact differentiable manifolds. Only slightly more argument gets the result for differentiable manifolds with the homotopy types of finite polyhedra.

Chapter Five

Section Four Analytic cocycles

We shall now give our final discussion of the way in which an analytic subvariety of a complex manifold defines a cohomology class.

We begin by defining the cohomology class of one oriented differentiable manifold imbedded in another

$$X \hookrightarrow M \quad .$$

Suppose that the codimension of X in M is i. The cohomology class of X in M, $[X]$, will be in $H^i(M, \mathbb{Z})$. We consider the normal bundle of this imbedding, $N \longrightarrow X$ (see Chapter Two, Section 2, and Milnor [23] for a discussion of normal bundles). N is a real oriented vector bundle of rank i. Let $N^x = N - \{\text{zero section}\}$. Since a neighborhood of the zero section in N is isomorphic to a neighborhood of X in M (see Milnor [23])

$$H^k(M, M-X, \mathbb{Z}) \xrightarrow{\simeq} H^k(N, N^x, \mathbb{Z})$$

for all k.

THEOREM W: There is a unique class $\tau \in H^i(N, N^x, \mathbb{Z})$ which localizes on each fiber to the generator of $H^i(\mathbb{R}^i, \mathbb{R}^i - \{0\})$ defined by the orientation.

We prove this first for the trivial normal bundle $\mathbb{R}^i \times X$, with the product orientation. There is the exact sequence

$$\cdots \longrightarrow H^j(\mathbb{R}^i \times X, \mathbb{Z}) \longrightarrow H^j(\mathbb{R}^i - \{0\} \times X, \mathbb{Z}) \longrightarrow H^{j+1}(\mathbb{R}^i \times X, \mathbb{R}^i - \{0\} \times X, \mathbb{Z}) \cdots$$

$$\longrightarrow H^{j+1}(\mathbb{R}^i \times X, \mathbb{Z}) \longrightarrow H^{j+1}(\mathbb{R}^i - \{0\} \times X, \mathbb{Z}) \longrightarrow \cdots$$

The map $H^j(\mathbb{R}^i \times X, \mathbb{Z}) \longrightarrow H^j(\mathbb{R}^i - \{0\} \times X, \mathbb{Z})$ is an isomorphism for $j < i - 1$, and is always injective. For $j = i-1$ the cokernel is an oriented copy of $\mathbb{Z}$, which implies the existence and uniqueness of τ in this special case.

Now we show the uniqueness of τ: If τ_1, τ_2 are both classes in $H^i(X \times \mathbb{R}^i, X \times \mathbb{R}^i - \{0\}, \mathbb{Z})$ which restrict to the generator on each fiber, let U be a maximal open subset of X on which $\tau_1 = \tau_2$. If U is not X, there is an open W not contained in U such that N is trivial over W. Let $N_U, N_W, N_{W \cap U}, N_{W \cup U}$ be the normal bundles. Then the Mayer-Victoris exact sequence

$$H^{\eta-1}(N_{W\cap U}, N^x_{W\cap U}) \longrightarrow H^{\eta}(N_{W\cup U}, N^x_{W\cup U}) \longrightarrow H^{\eta}(N_W, N^x_W) \oplus H^{\eta}(N_U, N^x_U) \longrightarrow H^{\eta}(N_{W\cap U}, N^x_{W\cap U})$$

together with the known result for $N_{W\cap U}, N_W$ shows that $\tau_1 = \tau_2$ on $W \cup U$, so U is all of X.

The same argument shows that τ exists.

The same argument also shows that $H^{\eta}(N, N^x, \mathbb{Z}) = 0$ for $\eta < i$.

The image of τ under

$$H^i(N, N^x, \mathbb{Z}) \xrightarrow{\simeq} H^i(M, M-X, \mathbb{Z}) \longrightarrow H^i(M, \mathbb{Z})$$

is called the fundamental class of X in M and denoted $[X]$.

We now know how to define the fundamental class of a smooth analytic subvariety of a complex manifold, since everything always has a natural orientation.

Now suppose that $X \hookrightarrow M$ is the inclusion of an analytic subvariety in a complex manifold, and that $X_1 = X_{sing}$, the singular locus of X is smooth. Then

$$H^{\eta}(M, M-X_1) = 0 \quad \text{for} \quad \eta < 2i+2$$

Then there is an exact sequence leading to an isomorphism

$$\underset{0}{H^{\eta}(M, M-X_1)} \longrightarrow H^{\eta}(M, M-X) \xrightarrow{\sim} H^{\eta}(M-X_1, M-X) \longrightarrow \underset{0}{H^{\eta+1}(M, M-X_1)}$$

if $\eta + 1 < 2i+2$. In particular,

$$H^{2i}(M, M-X) \xrightarrow{\sim} H^{2i}(M-X_1, M-X) \quad .$$

In case X_1 has singularities one excises these in turn, continuing until there are no singularities. In this way the cohomology class of an analytic subvariety is always defined.

Because of its intuitive appeal, we shall also give a definition of the cohomology class of an analytic subvariety in terms of deRham cohomology.

On a complex manifold of dimension n we denote

$H^i_{DR}(M, \mathbb{R})$ = group of closed real-valued C^{∞} differential forms, modulo exacts.

$H^i_{DR,c}(M,\mathbb{R})$ = group of closed real-valued C^∞ differential forms with compact supports, modulo exacts.

deRham's theorems give isomorphisms between $H^i_{DR}(M,\mathbb{R})$ and the real singular cohomology of M, and a duality between $H^i_{DR}(M,\mathbb{R})$ and $H^{2n-i}_{DR,c}(M,\mathbb{R})$, by $(\omega, \eta) \longrightarrow \int_M \omega \wedge \eta$.

If $X \hookrightarrow M$ is an analytic subvariety of pure codimension i, the $[X] \in H^{2i}_{DR}(M,\mathbb{R})$ will be defined as a functional on $H^{n-2i}_{DR,c}(M,\mathbb{R})$.

Denote by $A^j_c(M)$ the compactly supported differential forms on M of degree j. We shall see that for any $\varphi \in A^{2i}_c(M)$,

$$\int_{X-X_1} \varphi$$

is well-defined, and for $\eta \in A^{2i-1}_c(M)$,

$$\int_{X-X_1} d\eta = 0$$

where the integrations are always taken with the natural orientations on $X - X_1$.

Because φ has compact support we may assume that M is a polydisk in $\mathbb{C}^n$, with coordinates $(z_1, \ldots, z_n)$ and that the projection on the first n-k coordinates induces a branched covering, say of degree d, from X onto the polydisk in $\mathbb{C}^{n-k}$.

We set $\omega = \frac{\sqrt{-1}}{2} \Sigma^n_{j=1} dz_j \wedge d\bar{z}_j$ a 2-form in the polydisk. For any n-k linearly independent complex tangent vectors of type $(1,0)$, $t_1, \ldots, t_{n-k}$

$$\omega^k(t_1, \bar{t}_1, t_2, \bar{t}_2, \ldots, t_k, \bar{t}_{n-k}) > 0 \quad .$$

Thus one can assume there is $c > 0$ such that

$$|\varphi(t_1 \bar{t}_1, t_2 \bar{t}_2, \ldots, t_{n-k}, t_{n-k})| < c\omega^k(t_1, \bar{t}_1, \ldots, t_{n-k} \bar{t}_{n-k}) \quad .$$

There fore to show that the integral is well-defined it is enough to show that

$$\int_{X - X_1} \omega^k \qquad \text{is finite .}$$

Now if Y is any analytic subset of $X - X_1$, then Y has measure 0 in the measure defined by ω^k. Then for the purposes of measure theory, $X - X_1$ is a d-sheeted cover of $V(Z_n, Z_{n-1}, Z_{n-k+\Delta})$. Then

$$\int_{X-X_1} \omega^k \leq d \int_{V(Z_n, Zn-1, \ldots, Z_{n-k+\Delta})} \omega^k$$

and this last integral is certainly finite.

It only remains to see that $\int_{X-X_1} = 0$. This follows from the extended form of Stokes' theorem proved in Stolzenberg [33].

We shall not give a proof that all our definitions of cohomology classes coincide.

An element of $H^{2i}(M, \mathbb{Z})$ is called an <u>analytic cocycle</u> if it is in the group generated by the fundamental classes of analytic subvarieties. We shall use several facts about analytic cocycles, without proof. For more discussion see King [38] and the references given there.

If N is a complex manifold and $f: M \longrightarrow N$ is a holomorphic map then the f^* map in cohomology takes analytic cocycles into analytic cocycles.

The cup product of two analytic cocycles is again an analytic cocycle, so $H^\cdot(M, \mathbb{Z})$ contains an analytic cocycle subring.

If $f: M \longrightarrow N$ is a holomorphic map, and $Y \hookrightarrow N$ is an analytic subvariety of pure codimension i, then in case $f^{-1}(Y)$ has pure codimension i

$$[f^{-1}(Y)] = f^*[Y]$$

if $f^{-1}(Y)$ is counted with multiplicities.

We can now complete the proof of the theorem in Chapter Five, Section Two, that the Chern classes of the universal bundles on the Grassmannians are represented by Schubert subvarieties. On Grass(k,n) we had the global holomorphic sections $\varphi_1, \dots, \varphi_n$ and we claimed that $c_i(U_k)$ was represented by the analytic set where $\varphi_n, \varphi_{n-1}, \dots, \varphi_n-(k-i)$ are linearly dependent. We denote the complement of this set by W. On W there is an exact bundle sequence

$$0 \longrightarrow O_{hol}^{k-i-1} \longrightarrow U_k \longrightarrow Q \longrightarrow 0$$

where Q has fiber dimension $i-1$. Then $c_i(U_k) = 0$ restricted to W, so $c_i(U_k)$ is in the image of

$$H^{2i}(\text{Grass}(k,n), W) \longrightarrow H^{2i}(\text{Grass}(k,n)) \quad .$$

Now if we can see that $H^{2i}(\text{Grass}(k,n), W)$ is 1-dimensional, then we shall have the result up to constant multiples.

The singularities of the Schubert variety representing $c_i(U_k)$,

$$(\underbrace{n-k-1, \ldots, n-k-1}_{i \text{ places}}, n-k, \ldots, n-k)$$

are contained in the Schubert variety representing $c_{i+1}(U_k)$,

$$(\underbrace{n-k-1, n-k-1, \ldots, n-k-1}_{i-1 \text{ places}}, n-k, \ldots, n-k)$$

and $c_i(U_k) - c_{i+1}(U_k)$ is a complex manifold which is topologically a cell. Then

$$H^{2i}(\text{Grass}(k,n), W) \xrightarrow{\simeq} H^{2i}(\text{Grass}(k,n) - c_{i+1}(U_k), W)$$

and topologically the pair Grass $(k,n) - c_{i+1}(U_k), W$ is like the pair $\mathbb{C}^i, \mathbb{C}^i - \{0\}$. Then the cohomology group is one dimensional.

Now we know that on Grass (k,n) $c_i(U_k) = \eta_i[c_i(U_k)]$, and we know that $\eta_1 = 1$. Consider the flag manifold $F(k,n) \xrightarrow{f}$ Grass (k,n), where $f^*(U_k)$ is topologically a sum of line bundles, each of which is itself holomorphic. By looking at everything in terms of these line bundles, and using the fact that the cohomology of Grass (k,n) injects into the cohomology of $F(k,n)$, we can see that $\eta_i = 1$.

THEOREM X : Let the complex manifold M be either a projective variety or an affine variety. The Chern classes of holomorphic bundles on M are analytic cocycles.

Consider first the projective case. Given a holomorphic bundle $E \longrightarrow M$, there is a holomorphic line bundle L such that $E \otimes L$ is induced by a holomorphic map to some Grassmannian by arguments similar to those in Chapter Five, Section

One. This shows that the Chern classes of $E \otimes L$ are analytic cocycles, since they pull back from the Grassmannian. We shall see in a later chapter that the Chern classes of E are in the ring generated by the Chern classes of $E \otimes L$ and those of L.

In the affine case the cohomology vanishing theorems to be discussed in Chapter Eight will show that it is induced by a holomorphic map to a Grassmannian, and we get the same result.

Chapter Six

Section 1 K-theory and Bott periodicity

The results in this chapter will be purely topological. Our topological spaces will always be paracompact, locally compact Hausdorff spaces, having the homotopy type of a finite simplicial complex; we shall call such spaces nice spaces. To each nice space X we will associate a ring $K(X)$, which will classify the complex vector bundles on X up to a geometrically describable equivalence relation.

The Bott periodicity theorem gives an isomorphism

$$K(X) \otimes K(\mathbb{P}^1) \xrightarrow{\sim} K(X \times \mathbb{P}^1) \qquad .$$

This isomorphism will be the essential ingredient in the construction of a natural isomorphism

$$K(X) \otimes \mathbb{Q} \xrightarrow{\sim} H^{\cdot, \text{even}}(X, \mathbb{Q}) \qquad ,$$

which will show that, over $\mathbb{Q}$, the even-dimensional cohomology classes are combinations of the Chern classes of vector bundles.

We begin by describing the ring $K(X)$.

THEOREM A: <u>Let</u> X, Y <u>be nice spaces, and</u> $f_0, f_1 : X \longrightarrow Y$ <u>homotopic maps</u>. <u>Suppose</u> $E \longrightarrow Y$ <u>is a vector bundle</u>. <u>Then</u> $f_0^* E \xrightarrow{\sim} f_1^* E$.

First note that if E is a bundle on a nice space X, and $A \subset X$ is a closed subset, then any section of E over A can be extended to a section of E in a neighborhood of A. This is proved first in the case of a trivial bundle by the

Tietze extension theorem, then in the general case by a partition of unity.

Now let

$$F : X \times I \longrightarrow Y$$

be the homotopy and denote by g_t the composite map

$$X \times I \longrightarrow X \xrightarrow{f_t} Y$$

The bundles on $X \times I$ F^*E and $g_t^x E$ are isomorphic restricted to $X \times \{t\}$, and this defines a section of $\mathrm{Hom}(F^*E, g_t^* E)$ over $X \times \{t\}$, and thus over a neighborhood. The section will define an isomorphism on a neighborhood. Thus for every $t \in I$ there is a neighborhood U of t such that $F^*E \xrightarrow{\sim} g_t^* E$ on $X \times U$. This implies that $f_0^* E \xrightarrow{\sim} f_1^* E$.

On a nice space X the set of isomorphism classes of complex vector bundles

has an addition $(E, E') \longrightarrow E \oplus E'$ and a multiplication $(E, E') \longrightarrow E \otimes E'$

$K(X)$ is defined as the free abelian group generated by $\mathrm{Vect}(X)$ modulo the relations

$$[E \oplus E'] - ([E] + [E']) \qquad .$$

$K(X)$ has the property that any map

$$\gamma : \mathrm{Vect}(X) \longrightarrow G$$

with G a group, which is additive, factors uniquely as

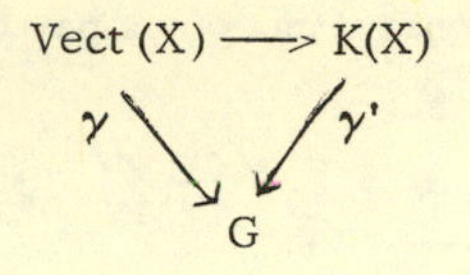

with γ' a group homomorphism.

Elements of $K(X)$ are called virtual bundles. Any element of $K(X)$ may be represented as $[E]-[E']$ with E, E' bundles on X. Two bundles E and F will define the same element of $K(X)$ just in case there is a third bundle G such that

$$E \oplus G \xrightarrow{\sim} F \oplus G \quad .$$

This condition is sufficient, for

$$[E]-[F] = [E \oplus G]-[F \oplus G] \quad .$$

To see that this is necessary, consider the monoid $\text{Vect}(X) \times \text{Vect}(X)/\text{diagonal}$. This monoid is in fact a group, and there is an additive map

$$\text{Vect}(X) \xrightarrow{\gamma} \text{Vect}(X) \times \text{Vect}(X)/\Delta$$

by

$$E \longrightarrow (E, 0)$$

and we will have $\gamma(E) = \gamma(F)$ only if there is G as above.

In fact, the induced map

$$\gamma' : K(X) \longrightarrow \text{Vect}(X) \times \text{Vect}(X)/\Delta$$

is an isomorphism.

We know that on a nice space an exact sequence

$$0 \longrightarrow E' \longrightarrow E \longrightarrow E'' \longrightarrow 0$$

of bundles always splits. Thus $K(X)$ could have been defined to have the universal property with respect to maps from $\mathrm{Vect}(X)$ to groups

$$\gamma : \mathrm{Vect}(X) \longrightarrow G$$

satisfying

$$\gamma(E) = \gamma(E') + \gamma(E'')$$

for exact sequences

$$0 \longrightarrow E' \longrightarrow E \longrightarrow E'' \longrightarrow 0 \qquad .$$

If F is another bundle then

$$0 \longrightarrow F \oplus E' \longrightarrow F \otimes E \longrightarrow F \otimes E'' \longrightarrow 0$$

is also exact, so $K(X)$ is actually a ring. A continuous map of nice spaces

$$f : X \longrightarrow Y$$

induces a ring homomorphism

$$f^* : K(Y) \longrightarrow K(X) \qquad .$$

On a nice space X with bundle E there is always a map

$$0 \longrightarrow K \longrightarrow I_d \longrightarrow E \longrightarrow 0$$

where I_d is the tirvial bundle of rank d. This would be immediate if X were compact, and it is true because X has a compact homotopy type. Splitting this sequence, we get $E \oplus K \xrightarrow{\sim} I_d$. It follows that E, F define the same elements of $K(X)$ just in case

$$E \oplus I_d \xrightarrow{\sim} F \oplus I_d$$

for some d. In other words, the relation defining $K(X)$ is <u>stable equivalence</u>.

Because any bundle on a nice space is a quotient of a trivial bundle, we have

THEOREM B: <u>On a nice space</u> X <u>there is a natural isomorphism</u>

$$\mathrm{Vect}_k(X) \xrightarrow{\sim} [X, \mathrm{Grass}\,(k,\infty)]$$

<u>where</u> $[\ ,\ ]$ <u>denotes homotopy classes of maps and</u> $\mathrm{Grass}\,(k,\infty)$ <u>is the limit of</u> $\mathrm{Grass}\,(k,n)$ <u>as</u> n <u>goes to infinity</u>. (Recall that $\mathrm{Grass}\,(k,n) \longrightarrow \mathrm{Grass}\,(k,n+1)$ as the Schubert variety representing the top Chern class of the universal quotient bundle U_k. We always pull back U_k.)

For example, $\mathrm{Grass}\,(1,\infty) = \mathbb{P}^\infty$ is the Eilenberg-MacLane space $K(\mathbb{Z},2)$ so $[X, \mathbb{P}^\infty] \xrightarrow{\sim} H^2(X,\mathbb{Z})$ which is the group of complex line bundles. For details, see Spanier [29].

K-theory began with Bott's computation of the homotopy groups of the complex general linear group. The natural map into the upper left corner

$$GL(n,\mathbb{C}) \longrightarrow GL(n+1,\mathbb{C})$$

induces a string of maps

$$\longrightarrow \pi_i(GL(n,\mathbb{C})) \longrightarrow \dots \longrightarrow \pi_i(GL(n+k),\mathbb{C}) \longrightarrow$$

It is a fact, to be discussed later, that these maps are eventually all isomorphisms. The resulting group

$$\pi_i(GL(\mathbb{C}))$$

is called the ith stable homotopy group of the general linear group, and this is what Bott computed.

We now relate this to vector bundles. If E is a vector bundle on a sphere S^n then E is trivial when restricted to $S^n - \{p\}$, p any point, because this space is contractible. Therefore any bundle on S^n arises by taking a trivial bundle on $S^n - \{\text{south pole}\}$ and one on $S^n - \{\text{north pole}\}$ and patching them along $S^n - \{\text{two points}\}$ with a map $S^n - \{\text{two points}\} \longrightarrow GL(k,\mathbb{C})$.

THEOREM C: Let X be a nice space, A and B nice subspaces such that $X = A \cup B$. Suppose given bundles E_A, E_B on A, B and a bundle isomorphism

$$g : E_A \xrightarrow{\sim} E_B$$

Then there is a bundle E on X and isomorphisms

$$E|_A \xrightarrow{i_A} E_A, \quad E|_B \xrightarrow{i_B} E_B$$

such that

$$g = i_B \circ i_A^{-1}|_{A \cap B}$$

The isomorphism class of E depends only on the homotopy class of g.

This is called the clutching construction for vector bundles.

To construct E we take the quotient of $E_A \amalg E_B$ by the obvious equivalence relation. This turns out to be a bundle. If g_0, g_1 are homotopic isomorphisms of $E_A|_{A\cap B}$, $E_B|_{A\cap B}$, then there is a bundle isomorphism

$$G : p^* E_A|_{I \times A\cap B} \xrightarrow{\sim} p^* E_B|_{I\times A\cap B}$$

which gives a clutching of p^*E_A and p^*E_B on $I \times X$, and induces g_0, g_1 on the ends. An argument like that in Theorem A shows that we get the same bundle on either end, namely the restriction of the total clutching.

Returning to S^n, an automorphism of the trivial bundle on $S^n - \{\text{two points}\}$ is given, up to homotopy, by an element of $[S^{n-1}, GL(k,\mathbb{C})]$, thus of $\pi_{n-1}(GL(k,\mathbb{C}))$. This identifies $\mathrm{Vect}_k(S^n)$ with $\pi_{n-1}(GL(k,\mathbb{C}))$.

We consider a slight generalization of this: For a nice space X, denote by $S_u(X)$ the unreduced suspension of X,

$$S_u(X) = X \times I/R$$

where the relation R identifies $X \times \{0\}$ to one point and $X \times \{1\}$ to another point.

THEOREM D : <u>There is a natural isomorphism</u>

$$\mathrm{Vect}_k(S_u(X)) \xrightarrow{\sim} [X, GL(k,\mathbb{C})] \quad .$$

Bott's original form of this theorem was

$$\pi_i(GL(\mathbb{C})) \xrightarrow{\sim} \pi_{i+2}(GL(\mathbb{C})) \quad .$$

One knows that $\pi_0(GL(\mathbb{C})) = 0$, $\pi_1(GL(\mathbb{C}) \xrightarrow{\sim} \mathbb{Z}$. Then for k big enough compared to n we have

$$\mathrm{Vect}_k(S^{2n}) \xrightarrow{\sim} \mathbb{Z}, \quad \mathrm{Vect}_k(S^{2n-1}) \xrightarrow{\sim} 0 \quad .$$

The explicit forms of those isomorphisms are of great interest and we shall discuss this later.

These lead to the isomorphism

$$K(S^{2n}) \longrightarrow \mathbb{Z}[X]/(X-1)^2 \quad , \quad K(S^{2n+1}) \xrightarrow{\sim} \mathbb{Z} \quad .$$

Again, the explicit form of the isomorphism is of great interest.

This also computes the homotopy groups of the infinite Grassmannians. For k big enough compared to i,

$$\pi_i(\mathrm{Grass}(k,\infty)) \xrightarrow{\sim} \pi_{i-1}(GL(\mathbb{C})) \quad .$$

For an elementary proof of the periodicity theorem in the form

$$K(X \times \mathbb{P}^1) \xrightarrow{\sim} K(X) \otimes K(\mathbb{P}^1)$$

see Atiyah [2] , or Bott [4] . Bott [4] also contains Bott's original proof. We shall see in the next section how to derive Bott's oroginal statement from the K-theory statement .

Chapter Six

Section 2 K-theory as a generalized cohomology theory

In this section we shall explore the formal aspects of K-theory by putting it in the setting of a general cohomology theory.

Let P denote the category of finite simplicial complexes. A general cohomology theory is a sequence of contravariant functors:

$$F^n : P \longrightarrow (\text{Abelian groups})$$

with $n \in \mathbf{Z}$, which are homotopy invariant and have the following property: For a polyhedral pair (X, Y) we define $F^n(X, Y)$ as the kernel of $F^n(X/Y) \longrightarrow F^n(pt)$, where X/Y denotes X with Y collapsed to a point. It is required that there be defined natural transformations $\delta : F^n(Y) \longrightarrow F^{n+1}(X, Y)$, called connecting homomorphisms, such that the long sequence

$$\longrightarrow F^n(Y) \xrightarrow{\delta} F^{n+1}(X, Y) \longrightarrow F^{n+1}(X) \longrightarrow F^{n+1}(Y) \xrightarrow{\delta} F^{n+2}(X, Y) \longrightarrow$$

is exact. Thus a cohomology theory is formally like ordinary simplicial cohomology.

We shall now build a cohomology theory out of K-theory. First functors K^n will be defined for $n \leq 0$ then Bott periodicity will be used to extend the definition to $n > 0$.

We use P^+ to denote the category of finite simplicial complexes with base points, and P^2 the category of simplicial pairs. For X in P^+, we define

$$\widetilde{K}(X) = \operatorname{Ker}(K(X) \xrightarrow{i^*} K(x_0))$$

where $i : x_0 \longrightarrow X$ is the inclusion of the base point. There is always a natural splitting

$$K(X) \xrightarrow{\simeq} \widetilde{K}(X) \oplus K(x_0) \qquad .$$

Defining a functor $P \longrightarrow P^+$ by $A \longmapsto A^+$ = disjoint union of A with a base point, we let $K^0(A) = \widetilde{K}(A^+)$. Of course $K^0(A)$ is just $K(A)$, which we are working into a formalism.

For X and Y in P^+ the <u>smash</u> <u>product</u> $X \wedge Y$ is defined by

$$X \wedge Y = X \times Y / (X \times y_0) \sqcup (x_0 \times Y)$$

with the natural base point. For X in P^+ a suspension, also in P^+, is defined as $S^1 \wedge X$ (we consider 1 as the base point of S^1). Then

$$\underbrace{S^1 \wedge \ldots \wedge S^1}_{n \text{ times}}$$

is isomorphic to S^n, and $S^n \wedge X$ is what we get by successively suspending X n times. We denote the suspension of X by $S(X)$, the n-fold suspension by $S^n(X)$.

For X in P we define

$$K^n(X) = \widetilde{K}(S^{-n}(X^+))$$

for $n < 0$.

THEOREM E: <u>For</u> $(X, Y) \in P^2$ <u>the sequence</u>

$$K^0(X, Y) \xrightarrow{j^*} K^0(X) \xrightarrow{i^*} K^0(Y)$$

<u>is exact</u>.

An element of the kernel of i^* may be represented as $[E] - [I_k]$ for some bundle E and some k. Since it is in the kernel we see that E is stably equivalent to I_k restricted to Y, so we may assume that E is trivial restricted to Y. The E can be lowered to $\widetilde{E}$ on X/Y, and $j^*([\widetilde{E}] - [I_k])$ is $[E] - [I_k]$.

Now we shall define the connecting homomorphisms.

THEOREM F: <u>For</u> $(X, Y) \in P^2$ <u>there is an infinite exact sequence to the left</u>

$$\longrightarrow K^{-2}(Y) \xrightarrow{\delta} K^{-1}(X, Y) \longrightarrow K^{-1}(X) \longrightarrow K^{-1}(Y) \xrightarrow{\delta} K^0(X, Y) \longrightarrow K^0(X) \longrightarrow K^0(Y) \longrightarrow 0$$

By taking suspensions, it will be enough to show that the sequences

(1) $K^{-1}(Y) \xrightarrow{\delta} K^0(X, Y) \longrightarrow K^0(X)$

(2) $K^{-1}(X) \longrightarrow K^{-1}(Y) \xrightarrow{\delta} K^0(X, Y)$

are both exact. Then the rest follows from the previous theorem.

To define δ we consider the cone over $X \in P$,

$$CX = X \times I / 1 \times X \quad .$$

CX has a natural base point. Note that CX/X may be identified with the unreduced suspension of X, if X is identified in CX with $0 \times X$.

The suspension S^1X is obtained from S_nX by collapsing a copy of I to a point. Then the exact sequence of the preceding theorem shows that

$$K(S^1X) \xrightarrow{\sim} K(S_nX)$$

which shows in particular that $K^{-1}(X)$ = virtual bundles on S_nX of rank 0.

Since $X \cup_Y CY/X \xrightarrow{\sim} S_nY$, there is a natural isomorphism

$$K^0(X \cup_Y (Y,X) \xrightarrow{\sim} K^0(S^1Y) = K^{-1}(Y) \quad .$$

The induced map

$$\begin{array}{c} \downarrow \\ K^{-1}(Y) \xrightarrow{\sim} K^0(X \cup_Y (Y,X) \longrightarrow K^0(X,Y) \end{array}$$

is δ. Then there is a commutative diagram

$$\begin{array}{ccccc} K^0(X \cup_Y (Y,X) & \longrightarrow & K^0(X \cup_Y CY) & \longrightarrow & K^0(X) \\ \downarrow s & & \downarrow s & & \downarrow s \\ K^{-1}(Y) & \xrightarrow{\delta} & K^0(X,Y) & \longrightarrow & K^0(X) \end{array}$$

using the identification of $K^0(X,Y)$ with $K^0(X \cup_Y CY)$ obtained from

$$X \cup_Y CY/CY \xrightarrow{\sim} X/Y$$

and the fact that CY is contractible. Since the top of the diagram is exact, this proves (1).

To prove (2), consider the commutative diagram

$$\begin{array}{ccccc}
K^0(X \cup_Y C_1Y \cup_X C_2X,\ X \cup_Y C_1Y) & \longrightarrow & K^0(X \cup_Y C^1Y \cup_X C^2X) & \longrightarrow & K^0(X \cup C_1Y) \\
\downarrow S & & \downarrow S & & \downarrow S \\
K^{-1}(X) & \xrightarrow{\ \varphi\ } & K^{-1}(Y) & \xrightarrow{\ \delta\ } & K^0(X, Y)
\end{array}$$

Here the top row is exact

$$X \cup_Y C_1Y \cup_X C_2X$$

means

$$(X \cup_Y CY) \cup CX \qquad .$$

The isomorphisms of the vertical arrows are constructed as before, and all maps except φ, which is induced by the isomorphisms and the top row, are the natural ones. We want to compare φ with the natural map $K^{-1}(X) \longrightarrow K^{-1}(Y)$.

Consider the special case where $X = S^1$, Y = a small arc as pictured. The general case is clear from this.

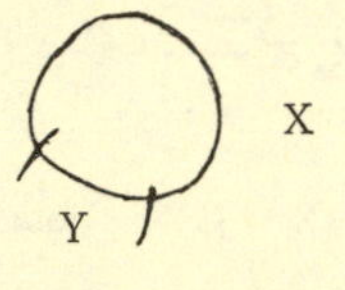

Then $X \cup_Y C_1Y \cup_X C_2X$ can be identified with a closed hemisphere of S^2, with a flap corresponding to C_1Y

C_1Y

The identification of $K^0(X \cup_Y C_1 Y \cup_X C_2 X)$ with $K^{-1}(Y)$ is obtained by observing that everything but the flap can contract to a point.

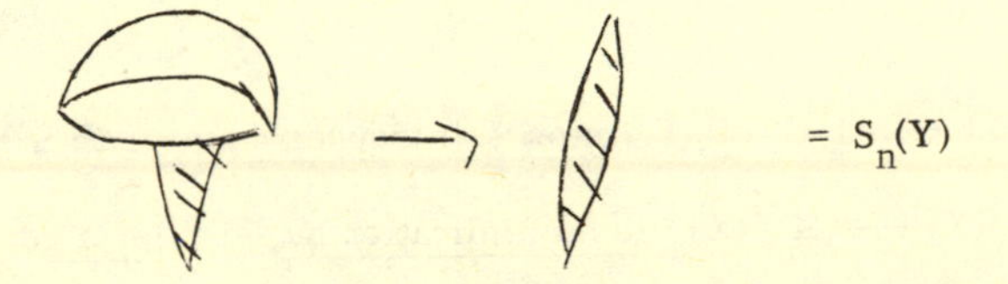

On the other hand, by contracting $X \cup_Y C_1 Y$ to a point, to get

$$K^0(X \cup_Y C^1 Y \cup_X C^2 X, X \cup_Y C^1 Y)$$

we get $S_n(X)$

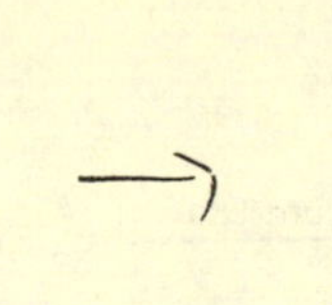
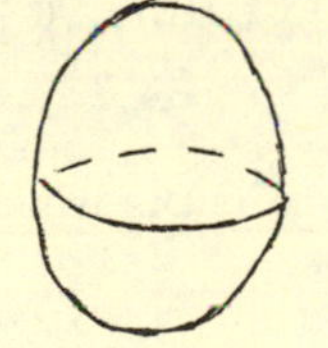

Imbedding everything in S^2

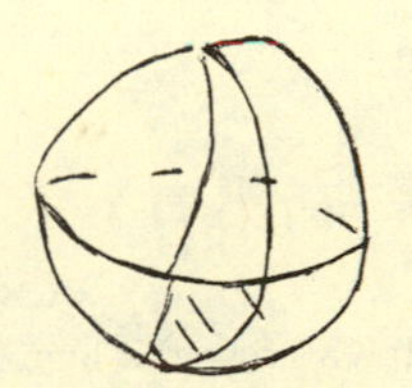

we see that in the first collapsing $S_n Y$ arises by collapsing the top hemisphere, and in the second case it is imbedded in $S_n X$ obtained by collapsing the bottom hemisphere. Hence we can factor φ as

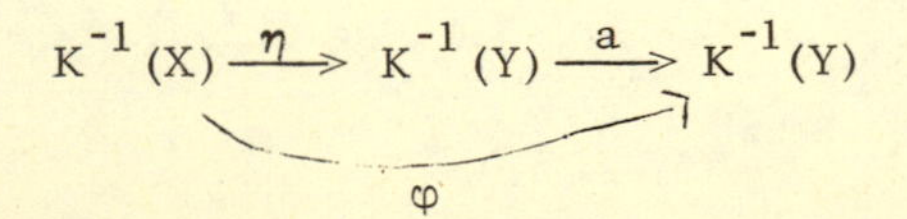

where η is the natural map and a is induced by the difference between collapsing the top and bottom hemispheres.

LEMMA : *The map* $t \longrightarrow 1-t$ *of* I *into itself induces a map* $\tilde{a} : S^1X \longrightarrow S^1X$. *The map this induces* $K^{-1}(X) \longrightarrow K^{-1}(X)$ *is multiplication by* -1, *for a general space* X .

This is a consequence of the following:

LEMMA : *For any map* $f : X \longrightarrow GL_{cn}, \mathbb{C})$, *let* E_f *denote the corresponding bundle over* S^1Y. *Then* $f \longrightarrow [E_f] - [I_n]$ *induces a group isomorphism*

$$\lim_{n \to \infty} [X, GL(n, \mathbb{C}] \longrightarrow \tilde{K}(S^1X)$$

where the group structure on the left is induced from that of $(GL_{cn}, \mathbb{C})$.

Since the map a corresponds in $[X, GL(n, \mathbb{C})]$ to $f \longmapsto \frac{1}{f}$, this will establish the first lemma.

We already have a bijection of sets

$$\lim_{n \to \infty} [X, GL(n, \mathbb{C})] \xrightarrow{\sim} \tilde{K}(S^1X) \qquad .$$

The fact that this is a group isomorphism follows from the homotopy equivalence of the two maps

$$GL(n) \times GL(n) \longrightarrow GL(2n)$$

given by

$$(A,B) \longmapsto \begin{pmatrix} A & 0 \\ 0 & B \end{pmatrix}$$

$$(A,B) \longmapsto \begin{pmatrix} AB & 0 \\ 0 & I \end{pmatrix}$$

with homotopy given by

$$P_t(A,B) = \begin{pmatrix} A & 0 \\ 0 & I \end{pmatrix} \begin{pmatrix} \cos t & \sin t \\ -\sin t & \cos t \end{pmatrix} \begin{pmatrix} I & 0 \\ 0 & B \end{pmatrix} \begin{pmatrix} \cos t & \sin t \\ \sin t & \cos t \end{pmatrix}$$

with $t \in [0, \frac{\pi}{2}]$.

COROLLARY G: If Y is a retract of X, there is an isomorphism for all n, $n \leq 0$

$$K^n(X) \xrightarrow{\sim} K^n(Y) \oplus K^n(X,Y) \quad .$$

This is a formal consequence of the existence of the connecting homomorphisms.

Applying this corollary to a product $A \times B$, where A, B are spaces with base points, one gets a formula for $\widetilde{K}(A \times B)$. For A is a retract of $A \times B$ and B is a retract of $A \times B/A$. Applying the corollary twice,

$$\widetilde{K}(A \times B) \xrightarrow{\sim} \widetilde{K}(A \wedge B) \oplus \widetilde{K}(A) \oplus \widetilde{K}(B)$$

and for general space X, Y

$$K^n(X \times Y) \xrightarrow{\sim} K^n(X) \oplus K^n(Y) \oplus K^n(X \wedge Y) \quad .$$

This shows that the kernel of the natural map $K^n(X \times Y) \longrightarrow K^n(X) \oplus K^n(Y)$ is identified with $K^n(X \wedge Y)$. Since the induced map

$$\tilde{K}(A) \otimes \tilde{K}(B) \longrightarrow \tilde{K}(A \times B) \longrightarrow \tilde{K}(A) \oplus \tilde{K}(B)$$

is zero, this leads to a pairing

$$\tilde{K}(A) \otimes \tilde{K}(B) \longrightarrow \tilde{K}(A \wedge B)$$

which induces a pairing for ordinary spaces X, Y

$$K^n(X) \otimes K^m(Y) \longrightarrow K^{n+m}(X \times Y)$$

(since $(X \times Y)^+ = X^+ \wedge Y^+$, and similarly for suspensions).

In particular, taking Y to be a point

$$K^n(\text{point}) \otimes K^m(X) \xrightarrow{\sim} K^{n+m}(X)$$

is given for all n $K^{-2}(\text{point}) = \tilde{K}(S^2)$, and the periodicity theorem gives

$$K^{-2}(\text{point}) \otimes K^0(X) \xrightarrow{\sim} K^{-2}(X) \quad .$$

By taking suspensions, we get

$$K^{-2}(\text{point}) \otimes K^n(X) \xrightarrow{\sim} K^{n-2}(X) \quad .$$

Now $K^{-2}(\text{point})$ is a free abelian group with one generator (it is $\pi_1(GL(\mathbb{C}))$, so $K^n(X) \xrightarrow{\sim} K^{n-2}(X)$ for all $n \leq 0$. In particular $\tilde{K}(S^n) = K^n(\text{point}) = \mathbb{Z}$ if n is even, $\tilde{K}(S^n) = 0$ if n is odd. This is the periodicity theorem in Bott's original formulation.

Because of the periodicity, the definition of $K^n(X)$ can be extended to $n > 0$ by $K^n(X) = K^{-n}(X)$. Then the sequence of functors K^n, $n \in \mathbb{Z}$, will have the formal properties of a cohomology theory.

We now want to compare K-theory with ordinary cohomology theory.

Chapter Seven

Section 1 The Chern character and obstruction theory

The goal of this chapter is to compare K-theory and ordinary cohomology. We shall give two different proofs of the basic theorem, one directly involving obstruction theory and the other a more formal proof involving a spectral sequence. At the end of this chapter we shall apply our results to study algebraic cocycles on a projective variety.

We shall study the Chern character, a map from Vect (X) to $H^{\cdot, \text{even}}(X, \mathbb{Q})$ which factors through K(X). The Chern class map from Vect (X) to $H^{\cdot, \text{even}}(X, \mathbb{Q})$ will not factor through K(X) because it is not additive. The Chern character, on the other hand, is additive and induces a ring homomorphism $K(X) \longrightarrow H^{\cdot, \text{even}}(X, \mathbb{Q})$.

We use the splitting principle to define this map. First, for line bundles, the map $L \longrightarrow 1 + c_1(L)$ is already additive, but not multiplicative. We set

$$\mathrm{ch}(L) = \exp(c_1(L)) = 1 + c_1(L) + \frac{c_1(L)^2}{2!} + \frac{c_1(L)^3}{3!} + \cdots \quad .$$

Then for an arbitrary bundle E we pass to the flag manifold $\pi : Y \longrightarrow X$ where $\pi^*(E) \xrightarrow{\sim} L_1 \oplus \cdots \oplus L_j$ is a sum of line bundles, and set

$$\mathrm{ch}(E) = \mathrm{ch}(L_1) + \cdots + \mathrm{ch}(L_j) \quad .$$

This will be expressible in the symmetric polynomials in $c_1(L_1), \ldots, c_1(L_j)$ and thus in the Chern classes of E. ch(E) so defined is in $H^{\cdot}(X, \mathbb{Q})$. The i-homogeneous part of ch(E) is

$$\frac{1}{i!}(c_1(L_1)^i + \ldots + c_1(L_d)^i)$$

while $c_i(E)$ is the ith symmetric polynomial in the $c_i(L_i)$. This shows that, over $\mathbb{Q}$, the $c_i(E)$ are polynomials in the components of $ch(E)$. $ch : Vect(X) \longrightarrow H^{\cdot, even}(X, \mathbb{Q})$ so defined gives an additive map, and one that satisfies $ch(E \otimes F) = ch(E)\, ch(F)$. Thus it induces a ring homomorphism $ch : K(X) \longrightarrow H^{\cdot, even}(X, \mathbb{Q})$, and thus a ring homomorphism

$$ch : K(X) \otimes \mathbb{Q} \longrightarrow H^{\cdot, even}(X, \mathbb{Q}) .$$

THEOREM A : For a nice space X, the map

$$ch : K(X) \otimes \mathbb{Q} \longrightarrow H^{\cdot, even}(X, \mathbb{Q})$$

is an isomorphism.

We first show that if the total Chern class of a bundle E is 1 then mE is stably trivial for some E. We then show that, up to multiples, bundles with given Chern classes can be constructed. We always assume that X is a polyhedron, and fix a triangulation.

We start by checking the case of a $2n$-sphere. In this case we shall prove the stronger result

$$ch : K(S^{2n}) \xrightarrow{\sim} H^{\cdot}(S^{2n}, \mathbb{Z})$$

where the map is given explicitly by

$$ch(E) = rank(E) + \frac{c_n(E)}{(n-1)!} .$$

In particular, $(n-1)!\,|\,c_n(E)$ for any bundle E on S^{2n}.

This will amount to showing that the isomorphism $\tilde{K}(S^{2n}) \xrightarrow{\sim} \mathbb{Z}$ is given explicitly by

$$[E] \longmapsto \frac{c_n(E)[S_{2n}]}{(n-1)!} .$$

This is clear on S^2 where every bundle is stably equivalent to a line bundle and $c_1 : \mathrm{Vect}(S^2) \xrightarrow{\sim} H^2(S^2, \mathbb{Z})$. The general case will follow by induction, using periodocity. Consider the diagram

$$\begin{array}{ccccccc} \tilde{K}(S^{2n} \times S^2) & = & \tilde{K}(S^{2n} \wedge S^2) & \oplus & \tilde{K}(S^2) & \oplus & \tilde{K}(S^{2n}) \\ \downarrow & & \downarrow & & \downarrow & & \downarrow \\ \tilde{H}^{\cdot}(S^{2n} \times S^2) & = & \tilde{H}^{\cdot}(S^{2n} \wedge S^2) & \oplus & \tilde{H}^{\cdot}(S^2) & \oplus & \tilde{H}^{\cdot}(S^{2n}) \end{array} .$$

The third and fourth vertical arrows are isomorphisms by induction, and the first is an isomorphism since $\tilde{K}(S^{2n} \times S^2) = \tilde{K}(S^{2n}) \otimes \tilde{K}(S^2)$, $\tilde{H}^{\cdot}(S^{2n} \times S^2) = \tilde{H}(S^{2n}) \otimes \tilde{H}^{\cdot}(S^2)$ and the Chern character commutes with these identifications. Since the indicated splittings are functorial, this gives a proof.

COROLLARY B : <u>The only spheres which could admit complex structures are S^2, S^4, and S^6.</u>

The proof will use the fact that on a compact complex manifold M^n, with tangent bundle T, $c_n(T)[M] = \chi(M)$. For a proof of this see Steenrod [30]. Combining this with the result just obtained we see that if S^{2n} has a complex structure then $(n-1)!\,|\,2$, so $n = 1, 2,$ or 3.

Of thesepossibilities it is known that S^2 has a complex structure. S^4 has no complex structure as can be seen from the Riemann-Roch theorem for complex surfaces. It is not known whether or not S^6 has a complex structure.

Before going into the proof of the theorem we need a few more remarks about the isomorphism $\tilde{K}(S^{2n}) \xrightarrow{\sim} \mathbb{Z}$. Suppose that E is a $\mathbb{C}^n$ bundle on S^{2n}, with global section φ such that φ is zero at a finite set of points on E. The <u>index</u> of an isolated zero p is computed by identifying a small neighborhood of p with B^{2n} (the ball in $\mathbb{R}^{2n}$) then using a trivialization of E on B^{2n} to make φ give a map

$$\varphi : B^{2n} - \{p\} \longrightarrow S^{2n-1} .$$

The degree of this map (with everything given an orientation induced from that of S^{2n}) is the index of φ at p. Then $c_n(E)[S_{2n}] = \sum_p$ index of φ at p. Again, a proof appears in Steenrod [30].

A $\mathbb{C}^n$-bundle on S^{2n} is given by a map

$$f : S^{2n-1} \longrightarrow GL(n, \mathbb{C}) ,$$

thus an element of $\pi_{2n-1}(GL(n, \mathbb{C}))$, say this is given by

$$x \longrightarrow (f^1(x), \ldots, f^n(x)) = \begin{pmatrix} f_1^1(x) & \cdot & f_1^n(x) \\ \vdots & \cdot & \vdots \\ f_n^1(x) & \cdot & f_n^n(x) \end{pmatrix} .$$

This induces a map $S^{2n-1} \longrightarrow S^{2n-1}$ by

$$x \longrightarrow \frac{f^1(x)}{\| f^1(x) \|} .$$

The degree of this map will be $c_n(E_f)[S^{2n}]$. To see this, note that $x \to f^1(x)$ can be extended to a map from the northern hemisphere into $\mathbb{C}^n$ by identifying the northern hemisphere with the closed 2n-ball and setting $f^1(x) = \|x\| f^1(\frac{x}{\|x\|})$ for $x \neq 0$, $f^1(0) = 0$. This will induce a section of E_f, trivialized as $(1, 0, \ldots, 0)$ on the lower hemisphere, f^1 on the upper. It will have an isolated zero only at the north pole, of index equal to the appropriate degree. In this way we have computed the isomorphism $\pi_{2n-1}(GL(N, \mathbb{C})) \xrightarrow{\sim} \mathbb{Z}$, for N big enough.

The proof of the theorem will now proceed by induction on the dimension of the polyhedron. Suppose then that E is a bundle on the polyhedron X , and that $c_q(E) = 0$ for $q > 0$. Consider first the case in which $2 \dim E = \dim X = 2k$. Then we may assume that E restricted to the 2k-1 skeleton of X is trivial $E|X^{2k-1} = I_k$. On the other hand, E restricted to any of the 2k disks ^{2k} which are attached to get X^{2k} is also trivial, and the comparison of these two trivializations on $\partial^{2k} = S^{2k-1}$ gives a map $\alpha_E : S^{2k-1} \to GL(k, \mathbb{C})$, thus an element of $\pi_{2k-1}(GL(k, \mathbb{C}))$. Setting $\eta_E(e^{2k})$ = the element of $\pi_{2k-1}(GL(k, \mathbb{C}))$ so defined every time a cell is attached, we get an element η_α of $C^{2k}(X, \pi_{2k-1}(GL(k, \mathbb{C}))) = C^{2k}(X, \mathbb{Z})$. Now $\eta_E(e^{2k}) = 0$ just in case the trivialization on X^{2k-1} extends across e^{2k} .

We will show that, at the cohomology level,

$$(k-1)! \, \eta_E = c_k(E)$$

so that our hypothesis will give $(k-1)! \, \eta$ cohomologous to zero. Then we can modify the given trivialization of E on X^{2k-1} to get a trivialization of $(k-1)! \, E$ extending over X^{2k} .

It follows from the previous description of $\pi_{2k-1}(GL(k,\mathbb{C}))$ that η_E, as an element of $C^{2k}(X,\mathbb{Z})$, is given by

$$\eta_E(e^{2k}) = \frac{\deg \alpha_E'}{(k-1)!} .$$

On the other hand, by the same argument as in the case of S^{2n}, $c^{2k} \longrightarrow \deg \alpha_E^1$ defines the cohomology class of $c_k(E)$ so that $(k-1)!\,\eta_E = c_k(E)$, in cohomology.

We can now dispense with our assumption that $2 \dim E = \dim X$. If $\dim E$ is arbitrary and dimension $X = 2k-1$ is odd, then we can in the same way define a cochain $\eta_E \in C^{k-1}(X, \pi_{2k}(GL(N,\mathbb{C})))$. Since $\pi_{2k}(GL(N,\mathbb{C})) = 0$ for N big enough, the trivialization always extends in this case.

If $\dim E$ is arbitrary and $\dim X = 2k$ is even, then E is induced by a map $\varphi: X \longrightarrow \mathrm{Grass}(r,n)$ for some $n, r = \dim E$. Outside of the Schubert variety representing $c_{k+1}(U_r)$ on the Grassmannian, U_r splits off a trivial bundle of rank $r-k$; since $\operatorname{codim}(c_{k+1}(U_r)) = k+1$, we may assume that $\varphi : X \longrightarrow \mathrm{Grass} - c_{k+1}(U_r)$, so $E = I_{r-k} \oplus E'$, E' a bundle of rank k. Now $c_k(E) = c_k(E')$, and the cochain $\eta_E \in C^{2k}(X, GL(r,\mathbb{C}))$ is the same as $\eta_{E'}$, if the trivializations are chosen properly. Thus one always has $(k-1)!\,\eta_E = c_k(E)$. Since the chern classes of E are those of E', we are reduced to the previous case. This completes the first part of the proof.

COROLLARY C : Suppose there is no torsion in $H^{\cdot,\mathrm{even}}(X,\mathbb{Z})$. Then $c_q(F)=0$, for $q>0$ implies that E is stably trivial.

Now we must prove that $\mathrm{ch} : K(X)\otimes\mathbb{Q} \longrightarrow H^{\cdot,\mathrm{even}}(X,\mathbb{Q})$ is surjective. We will show that, given $\eta\in H^{2n}(X,\mathbb{Z})$, there is $E\in \mathrm{Vect}(X)$ such that $c_q(E)=0$, $0<q<n$, $c_n(E)=m\cdot\eta$ for some m.

We first need some knowledge of the homotopy groups of Grass (k,n). It will be sufficient to know that

$$\pi_k(\mathrm{Grass}(r,2r)) = \begin{cases} \mathbb{Z} & \text{for } k \text{ even} \\ 0 & \text{for } k \text{ odd} \end{cases}.$$

for r big enough. For r big enough compared to k, one always has

$$\pi_k(\mathrm{Grass}(r,2r)) \xrightarrow{\sim} \mathrm{Vect}_r(S^k) \xrightarrow{\sim} \pi_{k-1}(GL(r,\mathbb{C})) .$$

(In fact, $\pi_k(\mathrm{Grass}(r,n)) \xrightarrow{\sim} \pi_{k-1}(GL(r,\mathbb{C}))$ if $n-r>k+1$. See Steenrod [30] for a proof.)

Now given $\eta \in C^{2k}(X, \mathbb{Z})$ we can construct, for big enough r, $E \in Vect_k(X^{2k})$ such that

$$(k-1)!\,\eta = c_k(E) \quad \text{over } X^{2k} \quad .$$

Take E to be trivial restricted to X^{2k-1}, and trivial on each attached $2k$-cell, patched together so that on $\partial(e^{2k})$, the induced element of $\pi_{2k-1}(GL(r, \mathbb{C}))$ is $\eta(e^{2k})$. Then, as we have seen before,

$$(k-1)!\,\eta = c_k(E)\,|\,X^{2k} \quad .$$

The bundle E is defined by a map

$$\psi : X^{2k} \longrightarrow \mathrm{Grass}\,(r, 2r) \quad ,$$

for r big enough. The obstruction to extending ψ over X^{2k+1} is an element of

$$C^{2k+1}(X^{2k+1}; \pi_{2k}(\mathrm{Grass}\,(r, 2r)))$$

say η_ψ. We will show that $\eta_\psi = \delta\eta$. Since $\eta_\psi(e^{2k+1})$ = the element of $\pi_{2k}(\mathrm{Grass}\,(2r, r))$ defined by $\psi\,|\,\partial e^{2k+1}$, $\eta_\psi(e^{2k+1}) = 0$ just in case ψ extends over this cell, consequently if η is a cocycle ψ will extend to X^{2k+1}.

η_ψ may be identified as $\eta_\psi(e^{2k+1})$ = element of $\widetilde{K}(S^{2k}) = E - I_r\,|_{\partial e^{2k+1}}$, which is clearly the obstruction to extending the map into the Grassmannian. On the other hand, we can interpret η by noting that $E - I_r$ is in the kernel of $K(X^{2k}) \longrightarrow K(X^{2k-1})$, thus identified with an element of $K(X^{2k}, X^{2k-1}) = \widetilde{K}(X^{2k}/X^{2k-1})$. Now $\eta(e^{2k})$ may be identified with the element of $\widetilde{K}(S^{2k})$ induced by the map

$$e^{2k}/\partial e^{2k} \longrightarrow X^{2k}/X^{2k-1}$$

pulling back $E - I_r$. Then $\eta_\psi = \delta\eta$.

This shows that, $\eta \in H^{2k}(X, \mathbb{Z})$ being given, we can find $E \in Vect_r(X^{2k+1})$ such that $c_q(E) = 0$, $0 < q < k$, $c_k(E) = (k-1)!\,\eta$. The obstruction to extending E over X^{2k+2} will be an element of

$$C^{2k+2}(X^{2k+2}, \pi_{2k+1}(\text{Grass}(r, 2r))$$

and since $\pi_{2k+1}(\text{Grass}) = 0$, we get this extension for free.

Now with the map $\psi : X^{2k+2} \longrightarrow \text{Grass}(r, 2r)$ there will be an obstruction to extending over X^{2k+3},

$$\omega \in C^{2k+3}(X^{2k+3}, \pi_{2k+2}(\text{Grass}(r, 2r))) .$$

Once again $\omega(e^{2k+3})$ = element of $\tilde{K}(S^{2k+2})$ defined by $E - I_r|_{\partial e^{2k+3}}$. To identify this note that we can split, by a previous argument,

$$E|X^{2k} = I_j \oplus \tilde{E} ,$$

where fiber dimension $\tilde{E} = k$. This splitting extends automatically to X^{2k+1}, $E|X^{2k+1} = I_j \oplus \tilde{E}$, and both $E, \tilde{E}$ extend to X^{2k+2}. Then $[E] - [\tilde{E}] \in K(X^{2k+2})$ is represented by a trivial bundle when restricted to X^{2k+1}, and so we know that the obstruction to extending $[E] - [\tilde{E}]$ to X^{2k+3} is $\delta\, ch([E] - [\tilde{E}])$. The obstruction to extending $[\tilde{E}]$ lies in $\pi_{2k+2}(\text{Grass}(k, \infty))$/ stability relation = $Vect_k(S^{2k+2})$/ stability relation. But the map

$$Vect_k(S^{2k+2}) \longrightarrow K(S^{2k+2})$$

is zero, since $K(S^{2k+2}) \longrightarrow H^{\cdot}(S^{2k+2})$ and the map is the Chern character, so

$\text{Vect}_k(S^{2k+2})$/stability is zero. Hence $[\tilde{E}]$ automatically extends so the obstruction to extending $[E]$ is $\delta\,\text{ch}([E]-[\tilde{E}])$. Since $\tilde{E}$ extends, $\delta\,\text{ch}_{2k+2}[\tilde{E}] = 0$ rationally, so the obstruction is (rationally) $\delta\,\text{ch}_{2k+2}(E) = \alpha$.

Our previous construction shows that we can find $E' \in \text{Vect}(X^{2k+2})$ such that $E' \mid X^{2k+1}$ is trivial, and $\text{ch}(E') = \alpha$. Then $\text{ch}([E]-[E']) = \eta$ in $H^{\cdot}(X^{2k+2}, \mathbb{Q})$, so $[E]-[E']$ will extend to X^{2k+3}, then for free over X^{2k+4}. The rational obstruction to the next extension will be $\delta\,\text{ch}_{2k+4}([E]-[E'])$ and we continue in the same way. This completes the proof of the theorem.

This theorem has an interesting corollary, which we will mention without proof. The result is due to Thom.

COROLLARY E : *Given a compact, oriented differentiable manifold* X, *and* $\eta \in H_{2k}(X, \mathbb{Z})$, *there exists a manifold* M_{2k}, *compact and oriented, and a map* $f : M_{2k} \longrightarrow X$ *such that* $f_*[M] = m \cdot \eta$, *for some* $m \in \mathbb{Z}$.

The idea of the proof is to first find a C^∞ bundle E on X such that $c_q(E)$ is the Poincaré dual of some multiple of η. Then for generic C^∞ sections of E, $g_1, \cdots, g_d$ where d = dimension of E, the subset Z where $g_1, g_2, \cdots, g_{d-q+1}$ are linearly dependent will be a sort of manifold with singularities and will represent the Poincaré dual of $c_q(E)$. Then we resolve the singularities of Z, which will be fairly simply, and set M_{2k} = the resolving manifold. $M_{2k} \longrightarrow X$ will be the natural map.

CHAPTER SEVEN

§2. The Atiyah-Hirzebruch Spectral Sequence

We shall now do much of the preceeding material over again, in a more formal setting.

THEOREM F: For each finite polyhedron X there is a spectral sequence $\{E_r, d_r\}$ with

$$E_2^{p,q} = H^p(X, K^q(\text{point}))$$

and

$$E_\infty^{p,q} = F_p K^{p+q}(X) \big/ F_{p+1} K^{p+q}(X)$$

where $F_p K^j(X)$ is defined as the kernel of the map

$$K^j(X) \longrightarrow K^j(X^{p-1}).$$

The spectral sequence is functorial in X, and the differentials of the sequence commute with suspension.

We define for each pair of integers p and q, with $p \geq q$

$$K(p,q) = K^{\bullet}(X^p, X^q) = \sum_{-\infty}^{\infty} K^n(X^p, X^q).$$

Then

(i) For each pair of pairs (p,q) and (p',q') with $p \geq p'$, $q \geq q'$, there is a natural map

$$K(p,q) \longrightarrow K(p',q')$$

(ii) For $p \geq q \geq r$ there is a map

$$\delta : K(q,r) \longrightarrow K(p,q)$$

such that the sequence

$$K(q,r) \xrightarrow{\delta} K(p,q) \longrightarrow K(p,r)$$

is exact.

Note that $K(p,q) = K(p',q)$ for $p, p' \geq \dim X$ and $K(p,q) = K(p,q')$ for $q, q' < 0$. Then we can set

$$K(\infty, q) = K(p,q) \quad \text{for } p \text{ big}$$

$$K(p, -\infty) = K(p, -1)$$

$$K(\infty, -\infty) = K^{\bullet}(X) = \sum_{-\infty}^{\infty} K^n(X).$$

The map δ is defined by

$$K^n(X^q, X^r) \xrightarrow{\alpha} K^n(X^q) \xrightarrow{p} K^{n+1}(X^p, X^q)$$

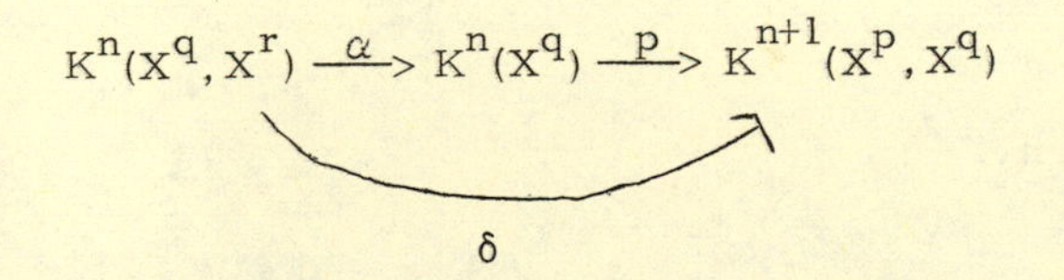

where α is the natural inclusion and β is the usual map of K-theory. The exactness mentioned in (ii) is then part of the exact sequence of a triple, which is a formal consequence of the cohomology properties of K-theory.

Because the data given satisfy (i) and (ii) there is, according to a purely algebraic theorem (see Cartan-Eilenberg [7]) a spectral sequence such that

$$E_r^{p,q} = Z_r^{p,q} \Big/ B_r^{p,q}$$

where

$$B_r^{p,q} = \mathrm{Ker}\left(K^{p+q}(p, p-1) \longrightarrow K^{p+q}(p, p-r)\right)$$

$$Z_r^{p,q} = \mathrm{Ker}\left(K^{p+q}(p, p-1) \xrightarrow{\delta} K^{p+q+1}(p+r-1, p)\right).$$

In particular

$$\begin{aligned} E_1^{p,q} &= \mathrm{Ker}(K^{p+q}(X^p, X^{p-1}) \xrightarrow{\delta} K^{p+q+1}(p,p) \\ &= K^{p+q}(X^p, X^{p-1}) \\ &= K^{p+q}\text{(a boquet of } p\text{-spheres, base point)} \\ &= \begin{cases} C^p(X, \mathbb{Z}) & \text{if } q \text{ is even} \\ 0 & \text{if } q \text{ is odd} \end{cases} \end{aligned}$$

as we see from Bott periodicity.

The differential

$$d_1\colon E_1^{p,q} \longrightarrow E_1^{p+1,q}$$

is given by

$$\delta: K^{p+q}(X^p, X^{p-1}) \longrightarrow K^{p+q+1}(X^{p+1}, X^p).$$

A computation like those we have done before shows that

$$d_1: C^p(X, \mathbb{Z}) \longrightarrow C^{p+1}(X, \mathbb{Z})$$

is the usual coboundary.

Note that for all r,

$$Z_r^{p,q} = K^{p+q}(X^p, X^{p-1}) = K^{p+q}(\text{boquet of } p\text{-spheres, point})$$

and this is 0 if q is odd.

Hence $E_r^{p,q} = 0$ for q odd, and

$$d_r: E_r^{p,q} \longrightarrow E_r^{p+r, q+r-1}$$

is zero for r even. In particular, $d_2 = 0$, so

$$E_3^{p,q} = E_2^{p,q} = H^p(X, K^q(\text{point})) = \begin{cases} H^p(X, \mathbb{Z}) & q \text{ odd} \\ 0 & q \text{ even} \end{cases}$$

Thus d_3 will be a holomorphism

$$d_3: H^p(X, \mathbb{Z}) \longrightarrow H^{p+3}(X, \mathbb{Z})$$

and d_3 is a natural transformation of functors

$$d_3: H^p(\ , \mathbb{Z}) \longrightarrow H^{p+3}(\ , \mathbb{Z})$$

defined on polyhedra, for all p. It is therefore a cohomology operation of type $(3, \mathbb{Z}, \mathbb{Z})$. d_5 will be a higher order cohomology operation

$$d_5 : \operatorname{Ker} d_3 | H^p \longrightarrow \operatorname{Coker} d_3 | H^{p+5}.$$

An important property of these operations is that they commute with suspension:

For any polyhedron X, there is a natural isomorphism

$$H^k(X, \mathbb{Z}) \longrightarrow H^{k+1}(S^1 X, \mathbb{Z})$$

induced by

$$H^k(X, \mathbb{Z}) \otimes H^1(S^1, \mathbb{Z}) \longrightarrow H^{k+1}(X \times S^1, \mathbb{Z}).$$

Bott periodicity similarly gives functorial isomorphisms in K-theory

$$K^n(X) \xrightarrow{\sim} K^{n+1}(S^2 X).$$

By suspending X and using these isomorphisms through the whole spectral sequence, we see that the diagram

$$\begin{array}{ccc} H^p(X) & \xrightarrow{d_3} & H^{p+3}(X) \\ \downarrow & & \downarrow \\ H^{p+1}(S^1 X) & \xrightarrow{d_3} & H^{p+4}(S^1 X) \end{array}$$

commutes, so d_3 is a stable cohomology operation -- that is, it commutes with suspension.

Our main result will be that if X is a complex manifold with the homotopy type of a polyhedron and $\eta \in H^p(X, \mathbb{Z})$ is an analytic cocycle, then $d_{2k+1}\eta = 0$ for all $k \geq 1$. Thus there are topological obstractions to a cocycle being analytic. To use this result we will want more information about the differentials.

THEOREM G: There is a unique non-zero stable cohomology operation of type $(3, \mathbb{Z}, \mathbb{Z})$, and some multiple of it is zero, so it always has its image in the torsion part of the cohomology group. Furthermore, all higher order stable cohomology operations defined on its kernel have finite order.

For proof of this, as well as a general discussion of cohomology operations, see Steenrod and Epstein [31], and Steenrod [32].

We shall use the following facts, discussed in these references: The unique stable cohomology operation of type $(3, \mathbb{Z}, \mathbb{Z})$ induces a unique stable cohomology operation of type $(3, \mathbb{Z}/2\mathbb{Z}, \mathbb{Z}/2\mathbb{Z})$. This is denoted Sq^3.

Actually, for each $i \geq 0$ there are stable cohomology operations, called the Steenrod squares

$$Sq^i : H^q(X, \mathbb{Z}/2\mathbb{Z}) \longrightarrow H^{q+i}(X, \mathbb{Z}/2\mathbb{Z}).$$

Later we shall need the factorization formula

$$Sq^k(uv) = \sum_{i+j=k} Sq^i u Sq^j v .$$

COROLLARY H: <u>There are isomorphisms</u>

$$K(X) \otimes \mathbb{Q} \xrightarrow{\sim} \sum_{n \text{ even}} H^n(X, \mathbb{Q})$$

$$K^1(X) \otimes \mathbb{Q} \longrightarrow \sum_{n \text{ odd}} H^n(X, \mathbb{Q}) .$$

Since all the differentials in our spectral sequence are killed by tensoring with $\mathbb{Q}$, we get isomorphisms

$$E_2^{p,q} \otimes \mathbb{Q} \xrightarrow{\sim} E_\infty^{p,q} \otimes \mathbb{Q} .$$

Now

$$E_\infty^{p,q} = F_p K^{p+q}(X)/F_{p+1}K^{p+q}(X)$$

$$F_p K^i(X) = \mathrm{Ker}(K^i(X) \longrightarrow K^i(X^{p-1})).$$

Then

$$\underset{\displaystyle E_2^{2p,0}}{\sum_{p=0}^{\infty} H^{2p}(X, \mathbb{Q})} \xrightarrow{\sim} \underset{\displaystyle E_\infty^{2p,0}}{\sum_{p=0}^{\infty} F_{2p}K(X)/F_{2p+1}K(X) \otimes \mathbb{Q}} .$$

But $F_{2p+1}K(X)/F_{2p+2}K(X)$ is an image of $K(X_{2p+1}, X_{2p}) = 0$. Hence

$$\sum_{p=0}^{\infty} H^{2p}(X, \mathbb{Q}) \xrightarrow{\sim} \sum_{p=0}^{\infty} F_{2p}K(X)/F_{2p+2}K(X) \otimes \mathbb{Q} \xrightarrow{\sim} K(X) \otimes \mathbb{Q}.$$

A similar argument works in the odd case.

Note that the isomorphisms obtained in this way are non-canonical. We shall compare them to the Chern character isomorphism.

First note that the isomorphism

$$E_1^{p,q} = K^{p+q}(X^p, X^{p-1}) \xrightarrow{\sim} C^p(X, \mathbb{Z})$$

for q even, is given by identifying

$$C^p(X, \mathbb{Z}) \xrightarrow{\sim} H^p(X^p, X^{p-1})$$

and the map

$$ch: \ K^p(X^p, X^{p-1}) \xrightarrow{\sim} H^p(X^p, X^{p-1})$$

gives the isomorphism explicitly. The map from the exact sequence of a triple gives the coboundary, or in other words d_1 gives the usual coboundary.

At the second level we have the isomorphism

$$E_2^{2p,0} \xrightarrow{\sim} H^{2p}(X, \mathbb{Z})$$

given explicitly as follows:

$$\xi \in Z_2^{2p,0} = \mathrm{Ker}(K^{2p}(X^{2p}, X^{2p-1}) \longrightarrow K(X^{2p+1}, X^{2p}))$$

is of the form $[E] - [I_r]$, where E is a $\mathbb{C}^r$ bundle on X^{2p}, trivial on X^{2p-1}, which extends over X^{2p+1} -- equivalent to $dch_p(E) = 0$, where

$$ch_p: K(X^{2p}, X^{2p-1}) \longrightarrow H^{2p}(X^{2p}, X^{2p-1}).$$

Thus ξ corresponds to $[ch_p \xi] \in H^{2p}(X^{2p+1}, \mathbb{Z}) = H^{2p}(X, \mathbb{Z})$. The element ξ will live forever in the spectral sequence just in case

$$\xi \in \mathrm{Ker}(K^{2p}(X^{2p}, X^{2p-1}) \longrightarrow K^{2p+1}(X, X^{2p}))$$

which is to say that ξ extends to all of X.

THEOREM I: α _in_ $H^{2p}(X, \mathbb{Z})$ _satisfies_ $d_{2k+1}\alpha = 0$ _for all_ k _just in case there is_ ξ _in_ $K(X)$, _trivial on_ X^{2p-1}, _such that_

$$ch(\xi) = \alpha + \text{higher order classes.}$$

Now we shall compute d_3 in a specific example and show that it is non-trivial. We begin with the double suspension of $\mathbb{P}^2, S^2(\mathbb{P}^2)$. Then

$$H^i(S^2(\mathbb{P}^2), \mathbb{Z}/2\mathbb{Z}) = \mathbb{Z}/2\mathbb{Z}$$

if $i=0,4$, or 6 and 0 otherwise. Furthermore, the cohomology operation

$$Sq^2\colon\ H^4(S^2(\mathbb{P}^2), \mathbb{Z}/2\mathbb{Z}) \longrightarrow H^6(S^2\mathbb{P}^2), \mathbb{Z}/2\mathbb{Z})$$

is not zero (see Spanier [29], where this is computed explicitly). Now we will attach a 7-cell to get our space. Note that $S^2(\mathbb{P}^2) = A_6$ has a cell decomposition with $A_4 = S^4$, $A_5 = S^4$, the map

$$S^5 \xrightarrow{\varphi} S^4$$

to attach γ^6 is the double suspension of the Hopf map, $S^3 \to S^2$.

If $g\colon S^5 \to S^5$ has degree 2, then the induced map

$$\varphi \circ g\colon\ S^5 \longrightarrow A_6$$

will be homotopically trivial, since $\pi_5(S_4) = \mathbb{Z}/2\mathbb{Z}$ (see Spanier [29]). Then $\varphi \circ g$ induces a map, $h\colon S^6 \to A_6$, and we use this to attach e^7, and get A_7. The induced map

$$h\colon\ H^6(A_6, \mathbb{Z}/2\mathbb{Z}) \longrightarrow H^6(S^6, \mathbb{Z}/2\mathbb{Z})$$

will be trivial, so

$$H^7(A_7, \mathbb{Z}/2\mathbb{Z}) = \mathbb{Z}/2\mathbb{Z},$$

$$H^i(A_7, \mathbb{Z}/2\mathbb{Z}) \xrightarrow{\simeq} H^i(S^2(\mathbb{P}^2), \mathbb{Z}/2\mathbb{Z}), \quad i<7.$$

We need another topological fact: If α is a generator of $H^4(A_7, \mathbb{Z}/2\mathbb{Z})$, then

$$Sq^3\alpha = \beta_2 Sq^2\alpha$$

where

$$\beta_2 : H^6(A_7, \mathbb{Z}/2\mathbb{Z}) \longrightarrow H^7(A^7, \mathbb{Z}/2\mathbb{Z})$$

is the Bockstein homomorphism. A proof of the factorization

$$Sq^3 = \beta_2 Sq^2$$

appears in Spanier [29], exercise G in Chapter Five. Hence β_2 is the Bockstein homomorphism induced by

$$0 \longrightarrow \mathbb{Z}/2\mathbb{Z} \longrightarrow \mathbb{Z}/4\mathbb{Z} \longrightarrow \mathbb{Z}/3\mathbb{Z} \longrightarrow 0 .$$

Now to show that $Sq^3\alpha$ is not zero we need only show that β_2 is not zero.

Hence β_2 is computed as follows: If c is in $Z^6(A_7, \mathbb{Z})$, and dc is zero mod 2, then

$$dc \in Z^6(A_7, 2\mathbb{Z}/4\mathbb{Z})$$

will represent $\beta_2(c)$. Now

$$C^6(A_7, \mathbb{Z}) = \mathbb{Z} = C^2(A_7, \mathbb{Z})$$

and the map from C^6 to C^7 is given by $b \longrightarrow 2b$. Hence β_2 is not zero. Then $Sq^3\alpha \neq 0$.

Now we will also use α to denote a generator of $H^4(A, \mathbb{Z})$. There is a virtual vector bundle E on $A_6 = S^2(\mathbb{P}^2)$ such that

$$ch(E) = \alpha$$

since

$$S^2(\mathbb{P}^2) = A_6 \quad , \quad ch: K(A_6) \xrightarrow{\sim} H^{\cdot}(A_6).$$

If E extended over A_7, then we would have $Sq^3\alpha = 0$, since then there would be

$$p: A_7 \longrightarrow Grass(2n, n)$$

such that $\alpha \in Im(p^*)$. But $Sq^3 = 0$ on all Grassmannians, because there is no even cohomology on the Grassmannian. Since Sq^3 is functorial, this shows what we wanted.

The example then shows that not all the differentials in the Atiyah-Hirzebruch spectral sequence are zero. In particular, since we have a class in H^4 on a seven-dimensional space for our example, $d_3 \neq 0$.

It will be convenient to recast the conditions of this last theorem in a slightly different form. As usual, X is a finite polyhedron. Consider the diagram

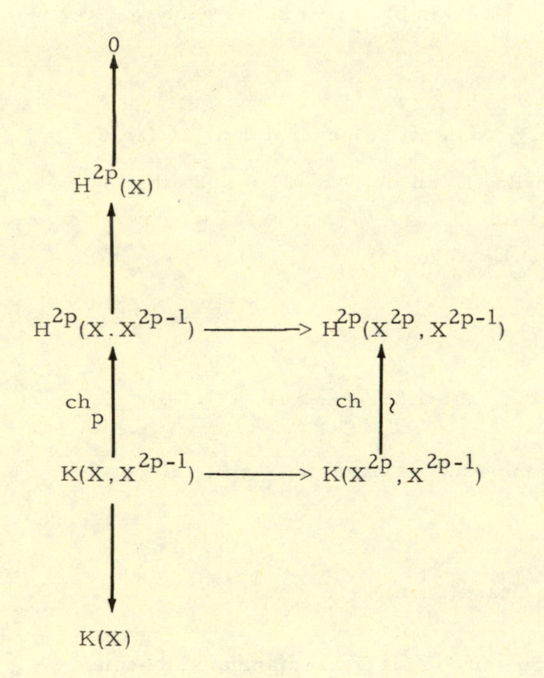

All cohomology is integral cohomology; since $H^{2p}(X, X^{2p-1})$ has no torsion,

$$ch_p : K(X, X^{2p-1}) \longrightarrow H^{2p}(X, X^{2p-1})$$

can be well defined; for any $\xi \in K(X, X^{2p-1})$

$$(p-1) \mid c_p(\xi) \in H^{2p}(X, X^{2p-1})$$

so it is well-defined. The diagram is commutative.

THEOREM J: $\alpha \in H^{2p}(X, \mathbb{Z})$ _lives forever in the spectral sequence just in case it lifts to_ $\tilde{\alpha} \in H^{2p}(X, X^{2p-1})$ _such that_

$$ch(\xi) = \alpha$$

for some $\xi \in K(X, X^{2p-1})$.

Now we will deduce conditions on an oriented real vector bundle of even dimension to be complex.

THEOREM K: _Let_ X _be a finite polyhedron, with complex vector bundle_ $E \to X$. _Let_ $B \to X$ _be an associated ball bundle, and_ $S \to X$ _the associated sphere bundle. Let_ d _be the fiber dimension of_ E. _Then the image of the Thom class_

$$H^{2d}(B, S, \mathbb{Z}) \longrightarrow H^{2d}(B, \mathbb{Z})$$

$$\gamma \longrightarrow [\gamma]$$

lives forever in the Atiyah-Hirzebruch spectral sequence.

We must construct $\xi \in K(B, B^{2d-1})$ such that $ch_p(\xi)$ maps to $[\gamma]$. The technique of constructing ξ is a difference construction: Given bundles A, B on X, and a bundle map $f\colon A \to B$ which is an isomorphism restricted to a subcomplex Y, one defines an element $d(A, B, f)$ of $K(X, Y)$ by a variant of the clutching construction. First glue together two copies of X along Y,

$$W = X_1 \cup_Y X_2 .$$

Then the isomorphism of A and B along Y allows us to clutch A on X_1 with B on X_2, to get $A \underset{f}{\cup} B$. The exact sequence

$$\begin{array}{ccccc} K(W/X_2) & \longrightarrow & K(W) & \longrightarrow & K(X_2) \\ \| & & & & \\ K(X,Y) & & & & \end{array}$$

splits because there is a projection $W \to X_2$, so we identify $K(X,Y)$ with the kernel of

$$K(W) \longrightarrow K(X_2).$$

Set

$$d(f,A,B) = \left[A \underset{f}{\cup} B\right] - [B].$$

For our purposes all this must be generalized. Suppose given a series of bundles and maps

$$A_1 \xrightarrow{f_1} A_2 \xrightarrow{f_2} \cdots \xrightarrow{f_{n-1}} A_n$$

such that

$$0 \longrightarrow A_1 \xrightarrow{f_1} A_2 \xrightarrow{f_2} \cdots \xrightarrow{f_{n-1}} A_n \longrightarrow 0$$

is exact along Y. We will construct $d(f_1, \ldots, f_n; A_1, \ldots, A_n)$ in $K(X,Y)$.

First note that, restricted to Y, there is a splitting

$$A_i \xrightarrow{\sim} B_i \oplus B_{i+1}$$

where

$$B_i = \mathrm{Ker}(f_i)$$

The B_j are defined only on Y. Now

$$\sum A_{2j-1} \xrightarrow{F} \sum A_{2j}$$

$$A_{2j-1} \xrightarrow{f_{2j-1}} A_{2j}$$

is an isomorphism, since each side is ΣB_i. We define $d(A_i, f_i)$ as $d(\Sigma A_{2j-1}, \Sigma A_{2j}, F)$.

A property of this construction which we will need is the following: If Y, Y^1 are subcomplexes of X and

$$L = [E_i, f_i]$$

is a sequence on X, acyclic over Y, and

$$L^1 = \left[E_j^1, f_j^1\right]$$

is a sequence on X_j acyclic over Y^1, then the complex $L \otimes L^1$ is acyclic over $Y \cup Y^1$, and $d(L \otimes L^1)$ is $d(L)d(L^1)$ under the pairing

$$K(X, Y) \otimes K(X, Y^1) \longrightarrow K(X, Y \cup Y^1).$$

The proof of this will be omitted.

Now to prove the theorem, we have the bundle $\Pi: E \to X$, and Π^*E has a tartological section e. This defines an exterior multiplication map

$$\wedge^i(\Pi^*E^*) \xrightarrow{e_i} \wedge^{i-1}(\Pi^*E^*)$$

and the sequence

$$0 \longrightarrow \wedge^d(\Pi^*E^*) \xrightarrow{e_d} \wedge^{d-1}(\Pi^*E^*) \longrightarrow \ldots \xrightarrow{e_2} \Pi^*E^* \longrightarrow \mathcal{O} \longrightarrow 0$$

is exact except on the zero section. Now restrict this to B. Since the zero section has codimension $2d$ in B, the inclusion $B_{2d-1} \longrightarrow B_{2d}$ may be moved away from the zero section by a small homotopy, so the above complex is homotopic to one exact on B_{2d-1}, and the difference construction gives an element γ of $K(B, B_{2d-1})$. We will show that the element of $H^{2d}(B, \mathbb{Z})$ which it defines is the image of the Thom class.

There is an isomorphism

$$H^{2d}(B, B_{2d-1}) \xrightarrow{\sim} H^{2d}(B, B^*)$$

which is natural. This serves to identify $ch_d(\gamma)$ naturally with an element of $H^{2d}(B, B^*)$, and we must show that it restricts to the generator on every fiber.

A fiber of B will be a unit ball $B^{2d} = \{|z| \leq 1 \text{ in } \mathbb{C}^d\}$ and the construction of γ will restrict to considering the sequence

$$0 \longrightarrow \Lambda^d(\mathbb{C}^{d*}) \xrightarrow{\text{int}_z} \Lambda^{d-1}(\mathbb{C}^{d*}) \xrightarrow{\text{int}_z} \cdots \xrightarrow{\text{int}_z} \mathbb{C}^{d*} \longrightarrow \mathbb{C} \longrightarrow 0.$$

In the special case $d = 1$ we have

$$0 \longrightarrow \mathbb{C} \xrightarrow{\text{mult by } z} \mathbb{C} \longrightarrow 0$$

is 1, and $\gamma(L) \in K(B^2, S^1) = \tilde{K}(S^2)$ is the Bott generator, as on sees from the description of the difference bundle.

In the general case we have a factorization of our complex

$$d(L) = d(L_1) \cdots d(L_d)$$

where L_i is

$$0 \longrightarrow \mathbb{C} \xrightarrow{z_i} \mathbb{C} \longrightarrow 0$$

defining

$$\gamma(L_i) \in K(B, z \in B, |z_i| > 1/2)$$

the Bott generator pulled back from the ith factor. Then d(L) is the image of the generator in the map

$$K(B^2, S^1) \otimes \cdots \otimes K(B^2, S^1) \longrightarrow K(B^{2d}, S^{2d-1})$$

and thus the Bott generator. This is what we wanted.

COROLLARY L: _Let_ M _be a complex manifold with the homotopy type of a finite polyhedron,_ X _a complex submanifold. Then_ $[X]$ _in_ $H^{2d}(M, \mathbb{Z})$ _lives forever in the Atiyah-Hirzebruch spectral sequence._

For the proof, let N be the normal bundle of X, with ball bundle N_B. Then

$$\tau \longmapsto [X]$$

under

$$H^{2d}(N_B, N_B^*, \mathbb{Z}) \xrightarrow{\sim} H^{2d}(M, M-X, \mathbb{Z}) \longrightarrow H^{2d}(M, \mathbb{Z})$$

and the theorem shows that all d_{2k-1} kill $[X]$.

Chapter Seven

Section 3 K-theory on algebraic varieties

Our immediate purpose is to prove that the analytic cocycles on a projective algebraic variety are generated by the Chern classes of analytic vector bundles.

Fix M as a projective algebraic manifold. We will define a ring $K_{alg}(M)$ analogous to $K_{top}(M)$ already investigated; start with the free abelian group generated by isomorphism classes of coherent algebraic sheaves on M, taking the quotient by the relations

$$[B] - ([A] + [C])$$

whenever $A \rightarrowtail B \twoheadrightarrow C$ is an exact sequence of such objects. $K_{alg}(M)$, like $K_{top}(M)$, is a group with a universal property.

A slightly different construction is involved in restricting ourselves to the locally free coherent algebraic sheaves; we follow the same procedure and get a group homomorphism $\tilde{K}_{alg}(M) \longrightarrow K_{alg}(M)$. $\tilde{K}_{alg}(M)$ may be thought of as algebraic vector bundles: it has a ring structure and there is a ring homomorphism $\tilde{K}_{alg}(M) \longrightarrow K_{top}(M)$.

THEOREM M: <u>The map</u>

$$\alpha : \tilde{K}_{alg}(M) \longrightarrow K_{alg}(M)$$

<u>is an isomorphism</u>.

We will construct an inverse $\gamma: K_{alg}(M) \longrightarrow \tilde{K}_{alg}(M)$. First note that any coherent algebraic sheaf F fits into a long exact sequence

$$0 \to E_d \to E_{d-1} \to \dots \to E_0 \to F \to 0$$

where the E_j locally free; everything up to the fact that the sequence terminates is a consequence of Theorems A and B ; the termination of the sequence is a consequence of Hilbert's syzygy theorem. (See Gunning-Rossi [13]) One can take d = dimension of M . This shows that α is surjective, since $[f] = \Sigma^d_{i=0} (-1)^i [E_i]$ in $K_{alg}(M)$.

Also define $\alpha(F) = \Sigma^d_{i=0}(-1)^i [E_i]$ in $\tilde{K}_{alg}(M)$. We must show that α does not depend on the resolution chosen. Suppose given two resolutions

$$E_{\cdot} \to F \to 0 , \quad E'_{\cdot} \to F \to 0$$

and a third, $E''_{\cdot} \to F \to 0$, together with epimorphisms $E''_{\cdot} \to E'_{\cdot}$, $E'_{\cdot} \to E_{\cdot}$. (an epimorphism in this context is a big commutative exact diagram

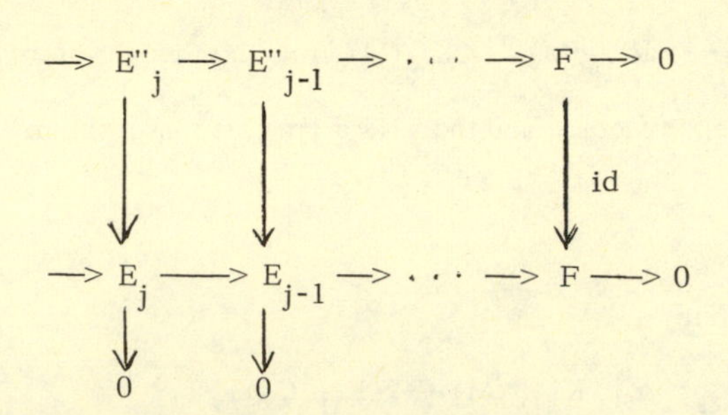

Then $\Sigma(-1)^i [E_i] = \Sigma(-1)^i [E''_i] = \Sigma(-1)^i [E'_i]$ in $\tilde{K}\,alg(M)$. Setting $K_i = \mathrm{Ker}(E''_i \to E_i)$, K_i will be locally free and the complex $0 \to K_d \to K_{d-1} \to \cdots \to K_0 \to 0$ will be acyclic. Then $\Sigma(-1)^i [K_i] = 0$, but $\Sigma_{(-1)^i} [E_j''] = \Sigma_{(-1)^i} [E_i] + \Sigma_{(-1)^i} [K_i]$.

Thus given $E'_\cdot$, $E''_\cdot$ we must construct $E''_\cdot$. This goes by induction on i . Suppose E''_i constructed we want to fill in the diagram

$$\begin{array}{ccc} 0 & & 0 \\ \uparrow & & \uparrow \\ E_{i+1} & \longrightarrow & E_i \\ & & \uparrow \\ E''_{i+1} & \longrightarrow & E''_i \\ & & \downarrow \\ E'_{i+1} & \longrightarrow & E'_i \\ \downarrow & & \downarrow \\ 0 & & 0 \end{array}$$

Let B_{i+1} Ker $(E_i \longrightarrow E_{i-j})$, and define B'_{i+1} , B''_{i+1} similarly.

Then we have a diagram

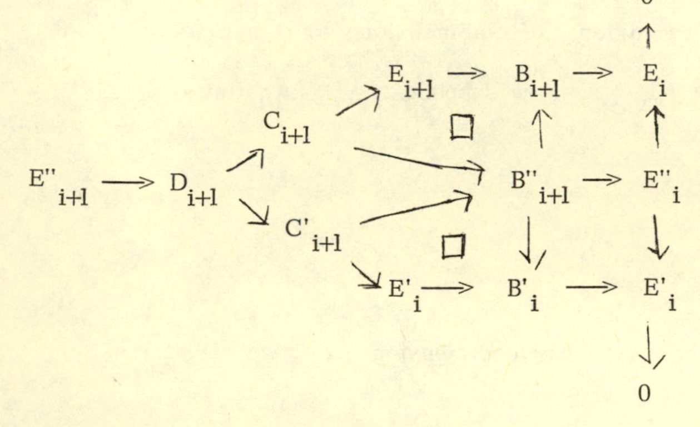

We may assume that $B''_{i+1} \to B_{i+1}$, $B''_{i+1} \to B'_i$ are surjective; if not, we modify E''_i to $E'_{i+1} \oplus E''_i \oplus E_{i+1}$, with trivial maps $E'_{i+1} \to E''_{i-1}$, $E_{i+1} \to E''_{i-1}$. With this assumption, we construct C_{i+1}, C'_{i+1}, D_{i+1}, and E''_{i+1} as follows:

Put $C_{i+1} = E_{i+1} \sqcup_{B_{i+1}} B''_{i+1}$ $(M \sqcup_P N = (m,n) \in M \times N :$

$f(m) = g(n)$, $f: M \to P$, $g : N \to P$ given)

with projection homomorphisms. Similarly, $C'_{i+j} = E'_{i+1} \sqcup_{B'_{i+1}} B''_{i+1}$.

Now take $D_{i+1} = C_{i+1} \times C''_{i+1}$, with projection homomorphisms.

Take E''_{i+1} locally free, $E''_{i+1} \to D_{i+1} \to 0$. Now the theorem follows.

This Lemma provides us with a ring structure on $K_{alg}(M)$ and a natural map

$$\beta : K_{alg}(M) \to K_{top}(M)$$

If $X \to M$ is an algebraic subvariety, of codimension d, then its class in $K_{top}(M)$ will be $\beta\,[O_X]$. Now suppose A is a finite polyhedron,

$$A \to M$$

is a homotopy equivalence. Since X has codimension d we may suppose that $A_{2d-1} \to M - X$.

Then we have the diagram

$$\begin{array}{ccc} H^{2d}(M, M-X) & \longrightarrow & H^{2d}(A, A_{2d-1}) \\ \downarrow & & \downarrow \\ H^{2d}(M) & \xrightarrow{\sim} & H^{2d}(A) \\ & & \downarrow \\ & & 0 \end{array} \qquad \mathbb{Z}\text{ - cohomology}$$

Now as an element of $K(A)$, $\beta([O_x])$ is 0 on A_{2d-1} ; for if

$$0 \to E_n \to E_{n-1} \to \ldots \to E_. \to O_X \to 0$$

is a resolution, then

$$0 \to E_n \to E_{n-1} \to \ldots \to E_. \to 0$$

is exact on $M-X$, so $\beta([O_X]) = \beta(\Sigma_{(-1)^i}\ [E_i]) = 0$ on A_{2d-1}.

Thus there is $ch_d(\beta[O_X]) \in H^{2d}(A, A_{2d-1})$. We want to show that the image of $ch_d(\beta[O_x])$ in $H^{2d}(A)$ corresponds to $[X]$ in $H^{2d}(M)$. The class corresponding to $ch_d(\beta[O_X])$ must restrict to 0 in $H^{2d}(M-X)$, so it comes from $H^{2d}(M, M-X)$. To show that it is $[X]$ let $x \in X$ be a regular point, with local coordinates $z_1, \ldots, z_n$ in a neighborhood $U = \|z\| < 1$ so $x = 0$, X is defined by $z_1 = \ldots = z_d = 0$.

A local computation exactly like that done at the end of the last section shows that $ch_j(\beta[O_X])$ locally gives the gnerator.

THEOREM N. _If_ M _is a projective algebraic manifold_, X _an analytic subvariety, then the cohomology class_ $[X]$ _is killed by all the_ d_k _of the Atiyah-Hirzebruch spectral sequence. In particular, considering_ $[X]$ _as a class mod 2,_ $Sq^3[X] = 0$.

THEOREM O. _If_ M _is a projective algebraic manifold, then the algebraic cocycles in_ $H^{\cdot}(M, \mathbb{Q})$ _are the image of the composite map_

$$K_{hol}(M) \otimes \mathbb{Q} \longrightarrow K(M) \otimes \mathbb{Q} \xrightarrow{ch} H^{\cdot}(M, \mathbb{Q}) .$$

This follows from the proof of the last theorem. The theorem is also true if

M is simply a complex manifold of finite homotopy type, X an analytic subvariety. The proof involves more technical difficulties--see Douady [39]. The original paper of Atiyah-Hirzebruch, to which we also refer, also has a formulation in the case where M does not have finite homotopy type. But we are most interested in the projective case. Consider the following

THEOREM P. _Let_ M _be a projective algebraic manifold_. _Then all of the torsion in_ $H^2(M, \mathbb{Z})$ _is algebraic_.

It will suffice to find a holomorphic line bundle L such that $c_1(L) = \alpha$, if α is a torsion class. The exact sequence of sheaves

$$0 \longrightarrow \mathbb{Z} \longrightarrow O_{hol} \xrightarrow{\exp 2\pi\sqrt{-1}} O^*_{hol} \longrightarrow 1$$

leads to

$$\begin{array}{c} H^1(M, O^*_{hol}) \xrightarrow{c_1} H^2(M, \mathbb{Z}) \longrightarrow H^2(M, O_{hol}). \\ \wr\!\uparrow \\ \mathrm{Vect}^1_{hol}(M) \end{array}$$

Since there is no torsion in $H^2(M, O_{hol})$ this proves the theorem.

We shall show that the generalization of this theorem is false: given $k > 1$, there is a projective algebraic manifold M and a torsion class in $H^{2k}(M, \mathbb{Z})$ which is not algebraic. All that is necessary is to find an M with a 2-torsion class $\alpha \in H^{2k}(M, \mathbb{Z})$ such that $Sq^3 \alpha \neq 0$. We will use a construction due to Godeaux and Serre.

THEOREM Q. _Let_ G _be a finite group,_ m _an integer_ ≥ 1. _There is a smooth projective variety_ M _which is_ $(m-1)$ _equivalent to the product of Eilenberg-Maclane spaces,_ $K(\mathbb{Z}, 2) \times K(G, 1)$. M _may be taken to have dimension_ m.

The construction begins with the following fact: If U is an open neighborhood of 0 in $\mathbb{C}^n$, and F a finite group of biholomorphic maps of U, then U/F has naturally the structure of an analytic space such that $\Gamma(U/F, O) = \Gamma(U, O)^F$. The reader can prove this himself.

We use this result to show that if M is a complex manifold, G a finite group of analytic automorphisms of M, then M/G has a natural analytic space structure. To define that structure locally, let $x \in M$ and $G_x = \{ g \in G : gx = x \}$. Let U' be a neighborhood of x, $U = \bigcup_{g \in G_x} gU'$. Then $G_x U' = U'$, and M/G looks locally like U'/G_x. But we have seen how U/G_x has an analytic structure.

If M is a projective variety then M/G will also be a projective variety: Let $L \longrightarrow M$ be a holomorphic line bundle which gives a projective imbedding. By passing to $\bigotimes_{g \in G} g^*L$ we may assume that L is invariant under G, so that it defines a holomorphic line bundle L' on $M/G = W$. Now if F is a coherent sheaf on W, we claim that $H^q(W, F \otimes L'^{\otimes d}) = 0$, for $q > 0$, d big enough. To prove this, take a covering $\{U_\alpha\}$ of W such that $\pi^{-1}(U_\alpha)$ on M is always cohomologically trivial. See Gunning-Rossi [13]. $\eta \in Z^q(\{u_\alpha\}, F \otimes L'^{\otimes d})$

defines $\pi^*\eta \in Z^q(\pi^{-1}\{u_\alpha\}, \pi^* F \otimes L^{\otimes d})$, and for d big enough, $q > 0$ there is $\alpha \in C^{q-1}(\{\pi^{-1} u_\alpha\}, \pi^* F \otimes L^{\otimes d})$ such that $\delta\alpha = \pi^*\eta$. Setting $\alpha' = \frac{1}{\#(G)} (\sum_{g \in G} g^*\alpha)$, $\alpha' \in C^{q-1}(\{u_\alpha\}, F \otimes L'^{\otimes d})$, $\delta\alpha = n$. This proves the vanishing theorem.

Now given a finite group G, it has a representation in $GL(N+1, \mathbb{C})$, for N big enough, such that the induced representation in $PGL(N+1, \mathbb{C})$ is faithful. Then we can construct a variety $\mathbb{P}^N/G$, with a projective imbedding.

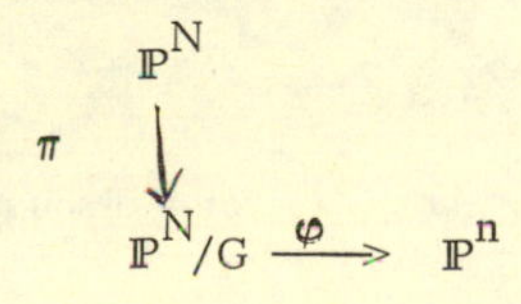

Our previous construction showed that we can pick the imbedding of φ so that $\pi^{-1}\varphi^{-1}$ (hyperplane section) = sum of $\#(G)$ hypersurfaces in $\mathbb{P}^N$.

Let $S = \pi$ (Fix G) be the image of the fixed point set of G. S is algebraic. For a given G, and m, we can always arrange things so that codimension S in $\mathbb{P}^n/G$ is $> m$. We could do this by taking a direct sum of several faithful representations of G in $GL(N+1)$ to get one in $GL(m(N+1))$.

Let $L \subset \mathbb{P}^n$ be a linear subvariety of dimension $N+m$. For generic L,

$$L \cap S = \varphi$$

$$L \cap \mathbb{P}^N/G \text{ is nonsingular, of dimension } m$$

(note $\mathbb{P}^n/G - S$ is smooth.)

so $L \cap \mathbb{P}^n/G$ is a smooth variety of dimension m. Also $\pi^{-1}(L \cap \mathbb{P}^N/G) = X$ is a smooth variety of dimension m. X will be invariant under the operation of G on $\mathbb{P}^N$, and it will be the intersection of $N-m$ hypersurfaces in $\mathbb{P}^N$.

It follows from the Lefschetz theorem that the map

$$X \xrightarrow{\ell} \mathbb{P}^N$$

in an $(m-1)$ equivalence, that is, that

$$\pi_i(X) \longrightarrow \pi_i(\mathbb{P}^N)$$

is an isomorphism for $i \leq m-1$. (See Milnor [29] for a discussion of the Lefschetz theorem. Using the map $\mathbb{P}^n \longrightarrow \mathbb{P}^\infty = K(\mathbb{Z}, 2)$, and the fact that $\pi_i(\mathbb{P}^n) = 0$, for $0 < i < 2n$, we see that there is an $m-1$ equivalence

$$X \xrightarrow{\ell} K(\mathbb{Z}, 2) .$$

Now we will show that the variety $M = X/G$ is $(m-1)$ equivalent to $K(\mathbb{Z}, 2) \times K(G, 1)$. First note that $X \xrightarrow{\ell} K(\mathbb{Z}, 2)$ is induced by a map, an $m-1$ equivalence, $M \xrightarrow{\ell} K(\mathbb{Z}, 2)$; let the map from M be induced by L', the map from X be induced by $\pi^* L' = L$. Then we have a diagram

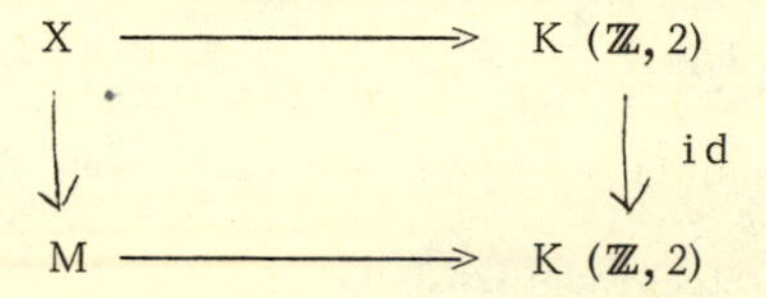

Over the space $K(G, 1) = B_G$ there is a universal principal G bundle, P_G, on which G operates freely with B_G as a quotient. For any space Y, there is an equivalence between

$$G\text{ - bundles over } Y \longleftrightarrow [Y, B_G]$$

The equivalence is obtained by pulling back P_G, which is a contractible space (one could take P_G = simplex with $\#(G)$ vertices). Then we get a diagram

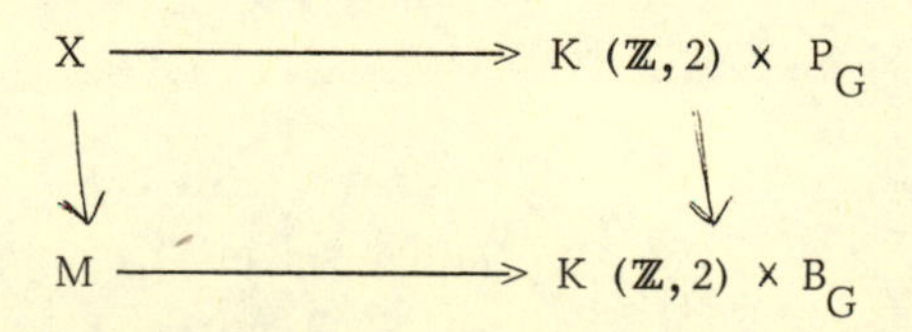

where both the top and bottom maps are $m-1$ equivalences. This proves the theorem.

Now one knows from the Whitehead theorem (see Spanier [29])

$$H^i(M, \mathbb{Z}) \xrightarrow{\sim} H^i(K(\mathbb{Z}, 2) \times B_G, \mathbb{Z})$$

for $i \leq m-1$. In particular, $H^i(B_G, \mathbb{Z})$ is a direct summand of $H^i(M, \mathbb{Z})$ for $i \leq m-1$.

To make our computation we shall use facts about the squaring operations, found in Spanier [29] or Steenrod [32].

Now take $G = \mathbb{Z}_2 \times \mathbb{Z}_2 \times \mathbb{Z}_2$. We have

$$B_{\mathbb{Z}_2} = \mathbb{P}^\infty(\mathbb{R}) = \lim_{n \to \infty} \mathbb{P}^n(\mathbb{R})_j$$

so $B_G = \mathbb{P}^\infty(\mathbb{R}) \times \mathbb{P}^\infty(\mathbb{R}) \times \mathbb{P}^\infty(\mathbb{R})_j$. For n odd, $H^\cdot(\mathbb{P}^n(\mathbb{R}), \mathbb{Z}_2)$ is $\mathbb{Z}_2[n] / n^{n+1}$, generated by an element of degree one. Now for large ℓ, $H^i(M, \mathbb{Z}_2)$ contains $H^i(\mathbb{P}^\ell(\mathbb{R}) \times \mathbb{P}^\ell(\mathbb{R}) \times \mathbb{P}^\ell(\mathbb{R}), \mathbb{Z}_2)$ as a direct summand. $H^\cdot(\mathbb{P}^\ell(\mathbb{R}) \times \mathbb{P}^\ell(\mathbb{R}) \times \mathbb{P}^\ell(\mathbb{R}), \mathbb{Z}_2)$ has 3 generators, n_1, n_2, n_3 in $H^1(\mathbb{P}^\ell(\mathbb{R}) \times \mathbb{P}^\ell(\mathbb{R}) \times \mathbb{P}^\ell(\mathbb{R}), \mathbb{Z}_2)$.

and

$$\begin{aligned} Sq^3(u_1u_2u_3^2) &= \sum_{i+j=3} Sq^i(u_1u_2)\, Sq^j(u_3^2) \\ &= Sq^3(u_1u_2) \cdot u_3^2 + Sq^2(u_1u_2) \cdot Sq^1(u_3^2) \\ &\quad + Sq^1(u_1u_2)\, Sq^2(u_3^2) + u_1u_2\, S_q^3(u_3^2) \end{aligned}$$

Now $Sq^3(u_1u_2) = Sq^3(u_3^2) = 0$, since $Sq^i(u) = 0$ for $i > \deg u$.

Also $Sq^1(u_1u_2)\, Sq^2(u_3^2) = u_1u_2u_3^4$

$$Sq^2(u_1u_2)\,Sq^1(u_3^2) = u_1^2u_2^2u_3$$

since $Sq^i(u) = u^i$ if $i = \deg u$. Thus $Sq^3(u_1u_2u_3^2) \neq 0$.

Since $u_1u_2u_3^2$ corresponds to a 2-torsion class in $H^4(M, \mathbb{Z})$, we see that there is a 2-torsion class in $H^4(M, \mathbb{Z})$ which is not complex analytic.

Chapter Eight

Section 1 Stein manifold theory

This chapter continues the study of vector bundles, now from an analytic point of view. Our results will tend to compare what can be done continuously on a complex manifold with what can be done holomorphically. For this purpose some study of Stein manifolds will be necessary.

A Stein manifold is a complex manifold with a strictly plurisubharmonic exhaustion function; that is, a complex manifold M will be Stein just in case there is a function $\tau : M \to \mathbb{R}^+$, which is C^2, such that $\tau^{-1}([0,\alpha])$ is compact for all $\alpha > 0$, and τ is strictly plurisubharmonic. This means that in any patch of M with local coordinates $z_1, \ldots, z_j$ the matrix $\frac{\partial^2 \tau}{\partial z_i \partial \bar{z}_j}$ will be positive definite. This is invariant under holomorphic change of coordinates.

A Stein manifold is a special type of Stein analytic space. These can be defined in a similar, but more complicated manner. This will not be discussed here. See Lelong [21] .

The two basic theorems of Stein manifold theory, which have already been mentioned in Chapter Two, are

THEOREM B. *If* F *is a coherent analytic sheaf on a Stein manifold* M , *then* $H^q(M,F) = 0$ *for* $q > 0$.

For the statement of theorem A we must know that an essentially unique topology can be defined on $H^0(M,F)$ for any complex manifold M and coherent analytic sheaf E . See Gunning-Rossi [13]. If is Stein with exhaustion function $\tau: M \to \mathbb{R}^+$, let

$M[r] = \tau^{-1}((0,r))$. It will also be a Stein manifold.

THEOREM A. The map $H^0(M,F) \to H^0(M[r],F)$ has dense image.

This is a generalization of the familiar Runge theorem in one variable. A consequence of this is the

THEOREM A'. On the Stein manifold M, F is generated by its global sections.

A simple consequence of Theorem B is the

THEOREM C. On a Stein manifold the natural map

$$\mathrm{Vect}^1_{hol}(M) \to \mathrm{Vect}^1_{top}(M)$$

is an isomorphism.

There is a natural isomorphism $\mathrm{Vect}^1_{top}(M) \xrightarrow{\sim} H^2(M, \mathbb{Z})$. The exact sequence $0 \to Z \xrightarrow{2\pi i} 0_{hol} \xrightarrow{\exp} 0^*_{hol} \to 1$ leads to the exact sequence

$$H^1(M,0) \to H^1(M,0^*) \to H^2(M,\mathbb{Z}) \to H^2(M,0)$$
$$\qquad\qquad \| $$
$$\qquad\qquad \mathrm{Vect}^1_{hol}$$

and the extreme terms vanish by Theorem B. This proves the theorem.

COROLLARY D. Every element of $H^2(M,\mathbb{Z})$ on a Stein manifold M is represented by a holomorphic divisor.

The divisor defined by a global section of a line bundle will define its Chern class.

We shall discuss the theorem of Grauert which generalizes this theorem to bundles of arbitrary dimensions. We shall also generalize the following

THEOREM F. *Let* M *be a Stein manifold. Denote by* $[M, \mathbb{C}^*]_{hol}$ *the class of holomorphic maps of* M *to* $\mathbb{C}^*$, *modulo holomorphic homotopy. Then the natural map* $[M, \mathbb{C}^*]_{hol} \to [M, \mathbb{C}^*]_{top}$ *is an isomorphism.*

A direct *holomorphic homotopy* between two holomorphic maps $f_0, f_1 : M \to \mathbb{C}^*$ consists of an holomorphic map $F : M \times \Delta(2) \to \mathbb{C}^*$ ($\Delta(2) = \{z \in \mathbb{C} : |z| < 2\}$) with $F(z,0) = f_0(z)$, $F(z,1) = f_1(z)$. Two maps f, g are holomorphically homotopic if there is a sequence $f = f_0, f_1, \ldots, f_k = g$ with f_i and f_{i+1} directly holomorphically homotopic.

To see the significance of this theorem, note that $\mathbb{C}^*$ is a $K(\mathbb{Z}, 1)$ so $[M, \mathbb{C}^*]_{top} = H^1(M, \mathbb{Z})$. On the other hand, there is a natural map $[M, \mathbb{C}^*]_{hol} \to H^1(M, \mathbb{Z})$ by $f \to \frac{df}{f}$, inducing an element of $H^1(M, \mathbb{Z})$ by the deRham isomorphism. We'll see that this is an isomorphism, so that all of the first cohomology can be realized holomorphically.

To prove the theorem we must use this result, which can be proven directly. Denote by Ω^p the sheaf of holomorphic p forms on M. Then there is the complex of sheaves.

$$0 \to \mathbb{C} \to \Omega^1 \xrightarrow{\partial} \Omega^2 \xrightarrow{\partial} \ldots \to \Omega^d \to 0$$

THEOREM G. *(holomorphic de Rham theorem) This sequence is exact.*

From this and Theorem B it follows that

$H^q(M, \mathbb{C}) \xrightarrow{\sim} \{\text{closed holomorphic } q\text{-forms}\}/\{\text{exact holomorphic } q\text{-forms}\}$

We'll do this after doing Theorem F.

An element of $[M, \mathbb{C}^*]_{top}$ defines an element w of $H^1(M, \mathbb{Z})$, thus of $H^1(M, \mathbb{C})$. Since w is integral, $\int_\gamma w \in \mathbb{Z}$ for any 1-cycle γ.

Now set, picking $z_0 \in M$,

$$f(z) = \exp\left(2\pi i \int_{z_0}^{z} w\right)$$

where the integral is over any path from z_0 to z. Then f is holomorphic, and defines the appropriate class. This shows that $[M, \mathbb{C}^*]_{hol} \to [M, \mathbb{C}^*]_{top}$ is surjective.

To show injectivity, suppose $\frac{df}{f} = 0$ in $H^1(M, \mathbb{C})$. Then $\log f = g$ exists on M, and defines $F : M \times \Delta(2) \to \mathbb{C}^*$ by $F(x, z) = \exp(zg(x))$. This shows $f = 0$ in $[M, \mathbb{C}^*]_{hol}$.

To prove the holomorphic de Rham theorem, let A^p denote the sheaf of differentiable, complex valued p-forms on M. Recall from Chapter Two the decomposition $A^p = A^{p,0} \oplus A^{p-1,1} \oplus \cdots \oplus A^{o,p}$. We have the commutative diagram of sheaf complexes

$$\begin{array}{ccccccccc} \to & A^p & \xrightarrow{d} & A^{p+1} & \xrightarrow{d} & A^{p+2} & \to \\ & \uparrow & & \uparrow & & \uparrow & \\ \to & \Omega^p & \xrightarrow{\partial} & \Omega^{p+1} & \xrightarrow{\partial} & \Omega^{p+2} & \to \end{array}$$

and the top row is known to be exact. Given φ a section of Ω^{p+1} over a neighborhood of p_1 with $\partial\varphi = 0$, we know that there is a section η of A^p over a perhaps smaller neighborhood with $d\eta = \varphi$. It will be enough to show that we can take $\eta \in A^{p,0}$, so $\bar{\partial}\eta = 0$ and $\eta \in \Omega^p$.

To make the modification, write

$$\eta = \Sigma_{j=0}^{k} \eta_{p-j,j} \quad , \quad \eta_{\ell,s} \in A^{\ell,s} \quad .$$

Now $\bar{\partial}\eta_{p-k,k} \in A^{p-k,k+1}$ is zero, so we use the $\bar{\partial}$ Poincare lemma (see Chapter Two) to conclude that on a slightly smaller neighborhood of p there is $\alpha \in A^{p-k,k-1}$ with $\bar{\partial}\alpha = \eta_{p-k,k}$. Set $\eta' = \eta - d\alpha$. Then

$$\eta' = \Sigma_{j=0}^{k-1} \eta^1_{p-j,j}$$

and $d\eta' = d\eta$. We continue in this way until we get an element of $A^{p,0}$. This proves the holomorphic de Rham theorem.

The theorem we shall discuss is that for Stein manifolds,

$$\mathrm{Vect}^k_{hol}(M) \xrightarrow{\sim} \mathrm{Vect}^k_{top}(M)$$

$$[M, GL(k,\mathbb{C})]_{hol} \xrightarrow{\sim} [M, GL(k,\mathbb{C})]_{top}$$

which shows that we can get $K^0(M)$ and $K^1(M)$ holomorphically. We know that $(K^0(M) \otimes \mathbb{C}) \oplus (K^1(M) \otimes \mathbb{C}) \xrightarrow{\sim} H^{\cdot}(M,\mathbb{C})$ and we have just seen how to get this holomorphically; more will be said about this later. We have just proved the theorem in the case $k = 1$.

Examples of Stein manifolds are

1. $\mathbb{C}^n$, with exhaustion function $\tau(z) = \|z\|^2$.
2. Any open complex manifold of dimension 1. (For a proof, see Narasihman [26]).
3. The product of Stein manifolds, with exhaustion function $\tau(x,y) = \tau_1(x) + \tau_2(y)$.
4. Any closed submanifold of a Stein manifold is Stein, using the restricted exhaustion function.

In particular, any affine, non-singular algebraic variety is a Stein manifold. The extent of the class of Stein manifolds is given by the

THEOREM H. (Bishop, Grauert, Narasimhan) Any Stein manifold of dimension n can be embedded as a closed submanifold in $\mathbb{C}^{2n+1}$.

For a proof see Hormander [17] or Gunning - Rossi [13]. After we have proved theorems A and B in special cases we shall use this result to get a proof in the general case.

As an example of a previous theorem, note that

$\mathbb{C}^* \times \mathbb{C}^* = \mathbb{P}^2 - V^+(x_1x_2x_3)$ is a Stein manifold, with the homotopy type of a torus, $S^1 \times S^1$. Hence $H^2(\mathbb{C}^* \times \mathbb{C}^*)$ is generated by an analytic curve--but not an algebraic curve, since any such curve can be pushed to infinity in $\mathbb{P}^2$.

Consider the holomorphic map from $\mathbb{C}^* \times \mathbb{C}^*$ into the Riemann surface

$\mathbb{C}(2\pi\mathbb{Z} + 2\pi\sqrt{-1}\,\mathbb{Z})$ defined by

$$(z_1, z_2) \longmapsto \frac{1}{\sqrt{-1}}\left(\log(z_1) + \log(z_2)\right)$$

This map defines a homotopy equivalence. A generator of the divisors on the Riemann surface is defined by a single point, so the analytic curve on $\mathbb{C}^x \times \mathbb{C}^x$ which we want is defined by

$$\frac{1}{\sqrt{-1}}\left(\log(z_1) + \log(z_2)\right) = \text{constant}$$

In the next section we shall give a proof that every holomorphic vector bundle on a polydisk is holomorphically trivial. The theorem that every topological vector has a unique holomorphic structure was proved by Grauert. His proof appears in Grauert [10, 11, 12]. A description of Grauert's proof appears in Cartan [6].

The main theorems of Stein manifold theory are proved in Hormander [17] and Gunning-Rossi [13].

CHAPTER EIGHT

§2. Holomorphic Vector Bundles on Polydisks.

Recall a construction of Chapter Five: If M is a differentiable manifold, $E \to M$ a differentiable vector bundle, the sheaf $V(E)$ on M is defined: First $\tilde{V}(E)$ is the subsheaf of the sheaf of tangent vectors on E, linear along the fibers, and $V(E)$ is the direct image of $\tilde{V}(E)$ under the projection. There is an exact sequence on M

$$0 \longrightarrow \underline{\mathrm{Hom}}(E,E) \longrightarrow V(E) \longrightarrow T(M) \longrightarrow 0.$$

To give a connection on E is to give a splitting of the sequence with a map

$$T(M) \xrightarrow{p} V(E)$$

Thus we know that connections always exist if M is paracompact.

For another proof of the existence of this splitting, consider the induced exact sequence of sheaves

$$0 \longrightarrow \underline{\mathrm{Hom}}(T(M),\underline{\mathrm{Hom}}(E,E)) \longrightarrow \underline{\mathrm{Hom}}(T(M),V(E))$$
$$\longrightarrow \underline{\mathrm{Hom}}(T(M),T(M)) \longrightarrow 0$$

To get a splitting is to get an element of $H^0(\underline{\mathrm{Hom}}(T(M,V(E)))$ mapping to the identity in $H^0(\underline{\mathrm{Hom}}(T(M),T(M))$. Since $H^1(\underline{\mathrm{Hom}}(T(M),\underline{\mathrm{Hom}}(E,E)) = 0$ (M is paracompact) we know we can do this.

Now if M is a complex manifold we can modify the discussion: Let E be a holomorphic vector bundle, $\tilde{V}(E)_{hol}$ the sheaf of holomorphic tangent vectors linear on fibers, and $V(E)_{hol}$ the direct image of $\tilde{V}(E)_{hol}$. Then we have an exact sequence of coherent analytic sheaves on M

$$0 \longrightarrow \underline{\mathrm{Hom}}(E,E)_{hol} \longrightarrow V(E)_{hol} \longrightarrow T(M)_{hol} \longrightarrow 0.$$

We define a holomorphic connection on M to be a splitting map $T(M)_{hol} \xrightarrow{p} V(E)_{hol}$. This induces a connection in the ordinary sense, a splitting $T(M)_{diff} \xrightarrow{p} V(E)_{diff}$. To say that the connection is holomorphic is equivalent to saying that Df is a holomorphic section of $E \otimes \Omega^1$ whenever f is a holomorphic section of E, all over some open set. Note that the curvature matrix of such a connection, with respect to a holomorphic frame, will be a matrix of holomorphic 2-forms.

THEOREM S: *If* $E \to M$ *is a holomorphic vector bundle on a Stein manifold then* E *admits a holomorphic connection. Consequently, the complex Chern classes* $c_i(E) \in H^{2i}(M, \mathbb{C})$ *are represented by holomorphic* $2i$-*forms*.

We repeat the discussion of the differentiable case, noting that $H^1(\underline{\mathrm{Hom}}(E,E)_{hol})T(M)_{hol}) = 0$ on a Stein manifold. The rest of the theorem follows from the interpretation of Chern classes as invariant polynomials in the curvature matrix, as given in Chapter Five.

Of course there is a less direct proof of the last part of this theorem from the holomorphic de Rham theorem. But now we have a nice way to construct $c_i(E)$ in $H^{2i}_{DR,hol}(M, \mathbb{C})$.

Our purpose now is to give a differential-geometric proof of the triviality of holomorphic vector bundles on $\mathbb{C}_n$. We must discuss flows of vector fields: If X is a differentiable manifold, and $\gamma_1, \ldots, \gamma_n$ are differentiable vector fields on X, we aim to find a differentiable action of $\mathbb{R}^n$ on X

$$\Phi : \mathbb{R}^n \times X \longrightarrow X$$

such that

$$\Phi(a + b, x) = \Phi(a, \Phi(b, x))$$

and

$$d\Phi_{(0,x)} : \frac{\partial}{\partial t_i} \longmapsto \gamma_i(x)$$

Of course Φ need not exist. Locally, the problem is, given $x \in X$ to find a neighborhood U_x of 0 in $\mathbb{R}^n$ and a map $\varphi_x : U_x \to X$ with $\varphi_x(0) = x$ and

$$d\varphi_{x(t_1, \ldots, t_n)} : \frac{\partial}{\partial t_i} \longmapsto \gamma_i(\varphi_x(t_1, \ldots, t_n))$$

According to the Frobenius theorem, there is at most one choice for φ_x, and it will exist on a small enough U_x just in case, using local coordinates $s_1, \ldots, s_m$ around x, so

$$\gamma_i(s_1,\ldots,s_m) = \sum_{j=1}^{m} \gamma_i^j(s_1,\ldots,s_m)\frac{\partial}{\partial s_j}$$

the relation

$$\sum_{\ell=1}^{m} \frac{\partial \gamma_i^j}{\partial s_\ell}\gamma_q^\ell - \frac{\partial \gamma_q^j}{\partial s_\ell}\gamma_i^\ell = 0$$

holds for any $q, i \in \{1,\ldots,n\}$, $j \in \{1,\ldots,m\}$. In other words, it is required that $[\gamma_i, \gamma_q] = 0$ for all i and q.

Φ will exist if we can take $U_x = \mathbb{R}^n$ for all x, in which case $\Phi(a,x) = \varphi_x(a)$. In any case, given vector fields with $[\gamma_i, \gamma_q] = 0$ we can set

$\epsilon(x) = \sup\{r > 0 \mid U_x$ contains an open ball of radius r about the origin$\}$.

Then $\epsilon(x) > 0$, and varies continuously with x. If $\epsilon(x) > \delta > 0$ identically on X then $\epsilon(x) = +\infty$: for given $a \in \mathbb{R}^n$ write $a = a_1 + \ldots + a_p$ with $|a_j| < g$ and set $\varphi_x(a) = \varphi_{\varphi_x(a_1)}(a_2 + \ldots + a_p)$ and so on. Thus we know that a global flow for

$$\gamma_i, \ldots, \gamma_n$$

with $[\gamma_i, \gamma_q] = 0$ always exists if X is compact.

Now suppose that X is a complex manifold and γ_1 is a holomorphic vector field on X. Then define γ_2 by $\gamma_2 = \sqrt{-1}\,\gamma_1$. This will satisfy $[\gamma_1, \gamma_2] = 0$, so given x in X we get a local flow

$$\varphi_x : U_x \longrightarrow X$$

U_x a neighborhood of 0 in $\mathbb{C}$. This map will be holomorphic, since it will be complex linear on the tangent spaces. Hence in case $\epsilon(x) > \delta > 0$ we get a holomorphic flow

$$\Phi : \mathbb{C} \times X \longrightarrow X$$

Similarly, given n holomorphic vector fields $\gamma_1, \ldots, \gamma_n$ with $[\gamma_i, \gamma_n] = 0$, set $\gamma_i^1 = \sqrt{-1}\, \gamma_i$, and we get in case $\epsilon(x) > \delta > 0$ a holomorphic flow

$$\Phi : \mathbb{C}^n \times X \longrightarrow X$$

Now let X be a differentiable manifold, $p: E \to X$ a differentiable vector bundle on X, with a connection. Suppose that $\gamma_1, \ldots, \gamma_n$ are vector fields on X, with lifts $\gamma_i \longmapsto \tilde{\gamma}_i \in V(E)$ Suppose that $[\gamma_i, \gamma_n] = 0$, and that $[\tilde{\gamma}_i, \tilde{\gamma}_n] = 0$ (this is not a consequence of the preceeding assumption).

LEMMA $\epsilon(y) = \epsilon(p(x))$ <u>for</u> $y \in E$

Given $\varphi_x : U_x \to X$ with appropriate properties, we want to lift it to $\tilde{\varphi}_x : U_x \to E$ such that $\tilde{\varphi}_x(0) = y$, $p \circ \tilde{\varphi}_x = \varphi_x$ and

$$d\tilde{\varphi}_{x(t_1, \ldots, t_n)} : \frac{\partial}{\partial t_i} \longmapsto \tilde{\gamma}_i(\tilde{\varphi}_x(t_1, \ldots, t_n))$$

There is an induced connection on $\varphi_x^*(E) \to U_x$, lifting $\frac{\partial}{\partial t_i}$ to $\frac{\tilde{\partial}}{\partial t_i}$ in $V(\varphi_x^*(E))$. It will be enough to find a section g of

$\varphi_x^*(E)$ such that $dg\left(\frac{\partial}{\partial t_i}\right) = \frac{\tilde{\partial}}{\partial t_i}$, $g(0)$ is y, and then set $\tilde{\varphi}_x$ equal to the composite

$$U_x \xrightarrow{g} \varphi_x^*(E) \longrightarrow E$$

Say the fiber of E is $\mathbb{C}^d$. Then $\varphi_x^*(E)$ is trivial and

$$V(\varphi_x^*(E)) = T(U_x) \oplus (\text{sheaf of maps, } U_x \longrightarrow g\ell(d, \mathbb{C}))$$

with the direct sum Lie algebra structure. This is locally the case on any manifold. Now

$$\frac{\tilde{\partial}}{\partial t_i}(t) = \frac{\partial}{\partial t_i} + A_i(t)$$

where

$$A_i : U_x \longrightarrow g\ell(d, \mathbb{C})$$

Because $[\tilde{\gamma}_i, \tilde{\gamma}_q] = 0$ we know that

$$\left[\frac{\tilde{\partial}}{\partial t_i}, \frac{\tilde{\partial}}{\partial t_q}\right] = 0$$

This means that $[A_i(t), A_q(t)] = 0$

Now set

$$g(t) = (t, \exp(t_1 A_1 + \ldots + t_n A_n) y)$$

This does what we wanted.

CORROLLARY U: Let M be a complex manifold, with a holomorphic connection in the holomorphic bundle $E \to M$. Then any holomorphic flow

$$\Phi : \mathbb{C} \times M \longrightarrow M$$

lifts to a holomorphic flow

$$\tilde{\Phi} : \mathbb{C} \times E \longrightarrow E.$$

If γ is a holomorphic vector field, lifting to $\tilde{\gamma}$ holomorphic on E, then $[\tilde{\gamma}, \sqrt{-1}\,\tilde{\gamma}] = 0$ automatically.

THEOREM V: Every holomorphic vector bundle on $\mathbb{C}^n$ is holomorphically trivial.

For $i \in \{1, \ldots, n\}$ there is a holomorphic flow

$$\Phi_i : \mathbb{C} \times \mathbb{C}^n \longrightarrow \mathbb{C}^n$$

given by

$$\Phi_i(w, z) = (z_1, \ldots, z_i + w, \ldots, z_n).$$

Thus Φ_i corresponds to $\partial/\partial z_i$. Fix a holomorphic connection in the holomorphic bundle E, to get

$$\tilde{\Phi}_i : \mathbb{C} \times E \longrightarrow E.$$

Now pick a basis $f_1, \ldots, f_d$ for the fiber of E over 0, and set

$$F_j(z) = \tilde{\varphi}_{n,z_n} \circ \tilde{\varphi}_{n-1,z_{n-1}} \circ \cdots \circ \tilde{\varphi}_{1,z_1}(f_j)$$

Then $F_1(z), \ldots, F_j(z)$ is a holomorphic frame for E.

We will make another application of the same technique. If M is a complex manifold, two holomorphic bundles E_0 and E_1 on are said to be _directly holomorphically homotopic_ if there is a holomorphic bundle E on $M \times D$ (D = unit disk in $\mathbb{C}$) such that

$$E \mid M \times \{0\} \xrightarrow{\sim} E_0, \quad E \mid M \times \{\tfrac{1}{3}\} \xrightarrow{\sim} E_1$$

Two bundles are _holomorphically homotopic_ if they can be connected by a chain of holomorphically homotopic bundles.

THEOREM W: _On a Stein manifold holomorphically homotopic bundles are isomorphic._

By integrating the holomorphic vector field $\partial/\partial z$ on $D_{1/2}$ (= disk of radius $1/2$) we get a restricted flow

$$\Phi : D_{1/2} \times (D_{1/2} \times M) \longrightarrow D \times M$$
$$(w,(z,x)) \longmapsto (w+z, x)$$

which we can lift to a linear-along-the-fibers flow

$$\tilde{\Phi} : D_{1/2} \times \left(E \mid_{D_{1/2} \times M}\right) \longrightarrow E$$

which induces an isomorphism $E_0 \xrightarrow{\sim} E_1$.

We remark that the triviality of any holomorphic vector bundle on the bounded polydisk is a simple consequence of theorem W.

COROLLARY X: *Let* U *be an open neighborhood of the closed polydisk in* $\mathbb{C}^n$ *given by* $|z_i| \leq 1$. *Let* $E \to U$ *be a holomorphic vector bundle. Then* E *is holomorphically trivial on the open polydisk given by* $|z_i| < 1$.

For some small $\epsilon > 0$ there is defined a holomorphic map

$$F : D_{1+\epsilon} \times \{|z_i| < 1\} \longrightarrow U$$

by

$$F(w, z) = (wz).$$

Then F_0 takes the open polydisk to a point and F_1 is the injection of the disk. Since $F_0^* E$ and $F_1^* E$ are holomorphically homotopic bundles, this proves the corollary.

Now an approximation argument will allow us to deal with open polydisks.

THEOREM Y: *Every holomorphic vector bundle on the polydisk* $\{|z_i| < 1\}$ *in* $\mathbb{C}^n$ *is holomorphically trivial.*

Pick an increasing sequence of positive numbers $\delta_1, \delta_2, \cdots, \delta_\ell, \cdots$ which converge to 1. Let $\Delta_\ell = \{|z_i| < \delta_\ell\}$. Suppose $E \to \{|z_i\} < 1\} = \Delta$ is our bundle, with fiber $\mathbb{C}^d$.

LEMMA (*Runge theorem for holomorphic maps to* $GL(d, \mathbb{C})$). *Let* $f: \overline{\Delta} \to GL(d, \mathbb{C})$ *be a map, defined and holomorphic on a* neighborhood of $\bar{\Delta}$. Given $\epsilon > 0$, there is a holomorphic map $g : \mathbb{C}^n \to GL(d, \mathbb{C})$ with $\sup_{\bar{\Delta}} |f-g| < \epsilon$.

We use the following facts: The set of holomorphic maps from $\bar{\Delta}$ to $GL(d, \mathbb{C})$, with topology defined by $\sup_{\bar{\Delta}} |f_1 - f_2|$, forms a topological group. This group is connected. For $GL(d, \mathbb{C})$ is connected, and $f_t(z) = f(tz)$ for $0 \leq t \leq 1$ defines a path between an element f of the group and a constant map.

For all A in $g\ell(d, \mathbb{C})$ the power series

$$\exp(A) = I + A + \frac{1}{2}A^2 + \frac{1}{6}A^2 + \ldots$$

converges to an element of $GL(d, \mathbb{C})$. If A and B commute then $\exp(A + B) = \exp(A)\exp(B)$. This exponential map is holomorphic. If $|I-C| < 1$ the series

$$\log C = (C-I) - \frac{1}{2}(C-I)^2 + \frac{1}{3}(C-I)^3 - \ldots$$

converges and $\exp \log C = C$. The log is holomorphic.

Now if $|f-I| < 1$ on a neighborhood of $\bar{\Delta}$, we can write $f = \exp \log f$ on that neighborhood. Looking at a power series expansion, there is a holomorphic $h: \mathbb{C}^n \to g\ell(d, \mathbb{C})$ which approximates $\log f$ up to ϵ^1 on $\bar{\Delta}$. Then we can make $\exp h$ approximate f very well on $\bar{\Delta}$, up to ϵ.

In general, we write

$$f = f_1 \circ f_2 \circ \ldots \circ f_m$$

where f_j is a holomorphic map from a neighborhood of $\overline{\Delta}$ to $GL(d, \mathbb{C})$. This is possible because a connected topological group is generated by any neighborhood of the identity. Approximating each f_j will get us what we want.

Returning to the proof of the theorem, we first pick $f_1^1, f_2^1, \dots, f_d^1$ holomorphic sections of E over $\overline{\Delta}_1$ which give a frame, then sections of E over $\overline{\Delta}_2$, $g_1^2, g_2^2, \dots, g_d^2$ which give a frame over $\overline{\Delta}_2$. Then there will be a holomorphic map h_1 into $GL(d, \mathbb{C})$, defined on a neighborhood of $\overline{\Delta}_1$, such that

$$\left(f_1^1, \dots, f_d^1\right) = h_1 \begin{pmatrix} g_1^2 \\ \vdots \\ f_d^2 \end{pmatrix}$$

where

$$\sup_{\overline{\Delta}_1} |I - k_1| < \frac{1}{2}, \quad |\det(k_1)| > \frac{1}{2}$$

In general, we find a holomorphic frame $f_1^\ell, \dots, f_d^\ell$ over $\overline{\Delta}_\ell$ with

$$\left(f_1^{\ell-1}, \dots, f_d^{\ell-1}\right) = k_{\ell-1} \begin{pmatrix} f_1^\ell \\ \vdots \\ f_d^\ell \end{pmatrix}$$

where

$$\sup |I - k_{\ell-1}| < \frac{1}{2^{\ell-1}}, \quad |\det(k_{\ell-1})| > \left(\frac{1}{2}\right)^{\frac{1}{2^{\ell-1}}}$$

This sequence of frames will converge to a holomorphic frame defined on all of Δ. This proves the theorem.

This same argument gives another proof of Theorem V.

Chapter Nine Concluding Remarks

In this chapter we shall give an overview of the problems dealt with in these notes. The motivating problem has been this: Given a projective algebraic variety, describe those cohomology classes which come from algebraic subvarieties. This is a problem originally studied by Picard, Poincaré, Lefschetz, and Hodge.

By the results of Chapter Four, our problem is the same as finding the analytic cocycles and by the results of Chapter Seven this is the same as finding which cohomology classes are represented by analytic vector bundles.

It is a result of Hodge that on a projective algebraic manifold, the decomposition

$$A^p = A^{p,0} \oplus \cdots \oplus A^{0,p}$$

of complex-valued differential forms descend naturally to cohomology with complex coefficients.

$$H^p(X, \mathbb{C}) = H^{p,0} \oplus \cdots H^{0,p}$$

We know that a holomorphic vector bundle always admits a connection with curvature of type (1,1). It follows from the results of Chapter Five that the p-th Chern class of a holomorphic vector bundle is represented by a differential form of type p, p. Thus the analytic cocycles in $H^{2p}(X, \mathbb{Z})$ are contained in the inverse image of $H^{p,p}$ under the map.

$$H^{2p}(X, \mathbb{Z}) \to H^{2p}(X, \mathbb{C})$$

For $p=1$ this is a necessary and sufficient condition, as was proved by Lefschetz--see Chern [8] for a proof.

The results of Chapter Seven show that there are torsion conditions on analytic cocycles for p bigger than one. But so far as anyone knows, every class in

$$H^{2p}(X, \mathbb{Q}) \cap H^{p,p}$$

is a rational multiple of an analytic cocycle.

Now on the affine variety we get by throwing away a hyperplane section, every rational cohomology class is represented by a continuous vector bundle, and hence by Grauert's theorem by an analytic vector bundle. Thus given a class of type p, p on a projective variety, we know that it becomes analytic when restricted to the complement of a hyperplane section. Thus we must introduce a singularity along the hyperplane section. Conjecturally this singularity is inessential for a class of type p, p. So far, the proof of Grauert's theorem does not allow one to keep track of the singularity introduced. But some results in this direction have been obtained by Griffiths and Cornalba.

BIBLIOGRAPHY

1. A. Altman and S. Kleiman, Introduction to Grothendieck Duality Theory, Springer Lecture Notes, Number 146.

2. M. Atiyah, K-Theory, Benjamin.

3. M. Atiyah and F. Hirzebruch, "Analytic Cycles on Complex Manifolds," Topology, vol. 1.

4. R. Bott, Lectures on K (X), Benjamin.

5. R. Bott, "On a Theorem of Lefschetz," Michigan J. of Math., vol. 6.

6. H. Cartan, "Espaces Fibres Analytiques," Symposium Internacional de Topologica Algebraica.

7. H. Cartan and S. Eilenberg, Homological Algebra, Princeton.

8. S. S. Chern, Complex Manifolds without Potential Theory, Van Nostrand.

9. R. Godement, Theorie des Faisceaux, Hermann.

10. H. Grauert, "Approximationssatze für Holomorphe Funktionen mit Werten in Komplexen Raum," Math. Ann., vol. 133.

11. ______ "Holomorphe Funktionen mit Werten in Komplexen Lieschen Gruppen," Math. Ann., vol. 133.

12. ______ "Analytischen Faserungen über Holomorph-Vollstandigen Raumen," Math. Ann., vol. 133.

13. R. Gunning and H. Rossi, Analytic Functions of Several Complex Variables, Prentice-Hall.

14. R. Gunning, Riemann Surfaces, Princeton Lecture Notes.

15. H. Hironaka, "Resolution of the Singularities of an Algebraic Variety over a Field of Characteristic Zero," I and II, Annals of Math., vol. 79.

16. N. Hicks, Notes on Differential Geometry, Van Nostrand.

17. L. Hormander, Introduction to Complex Analysis in Several Variables, Van Nostrand.

18. S. Kobayashi and K. Nomizu, Foundations of Differential Geometry, J. Wiley.

19. K. Kodaira and J. Morrow, Complex Manifolds, Holt, Rinehart, and Winston.

20. S. Lefschetz, Algebraic Geometry, Princeton.

21. P. Lelong, Fonctions Plurisousharmoniques et Formes Differentielles Positifs, Gordon and Breach.

22. H. Matsumara, Commutative Algebra, Benjamin.

23. J. Milnor, Differential Topology, Princeton Lecture Notes.

24. J. Milnor, Morse Theory, Princeton.

25. D. Mumford, Introduction to Algebraic Geometry, Harvard Notes.

26. R. Narasimhan, Introduction to the Theory of Analytic Spaces, Springer Lecture Notes, number 25.

27. J.P. Serre, "Faisceaux Algebrique Coherents," Annals of Math., vol. 61.

28. J.P. Serre, "Geometrie Algebrique et Geometrie Analytique," Ann. Ann. Inst. Fourier, vol. 6.

29. E. Spanier, Algebraic Topology, McGraw-Hill.

30. N. Steenrod, Fibre Bundles, Princeton.

31. N. Steenrod and D. Epstein, Cohomology Operations, Princeton.

32. N. Steenrod, "Cohomology Operations," Advances in Mathematics, vol. 9.

33. G. Stolzenberg, Volumes, Limits, and Extensions of Analytic Spaces, Springer Lecture Notes, number 19.

34. R. Swan, Sheaf Theory, Chicago Lecture Notes.

35. I. Safarevic, "Introduction to Algebraic Geometry," Russian Math. Surveys.

36. R. Bott, Morse Theory, Harvard Notes.

37. S. Kleiman, "Geometry on Grassmannians and Applications," in Publications I. H. E. S., vol. 36.

38. J. King, "The Currents Defined by Analytic Varieties," Acta Math., vol. 127.

39. A. Douady, Exposé in Sem. Bourbaki on analytic cycles.

40. R. Stong, Notes on Cobordism, Princeton Notes.

Library of Congress Cataloging in Publication Data

Griffiths, Phillip, 1938-
Topics in algebraic and analytic geometry; notes from a course of Phillip Griffiths.

(Mathematical notes, 13)
"Revised version of the notes taken from a class taught at Princeton University in 1971-1972."
Bibliography: p.
1. Geometry, Algebraic. 2. Geometry, Analytic. I. Adams, John, 1949- II. Title. III. Series: Mathematical notes (Princeton, N. J.), 13.
QA564.G66 516'.3 74-2968
ISBN 0-691-08151-4